Indoor Air Quality and HVAC Systems

David W. Bearg, P.E.
Life Energy Associates
Concord, Massachusetts

LEWIS PUBLISHERS
Boca Raton   Ann Arbor   London   Tokyo

Library of Congress Cataloging-in-Publication Data

Bearg, David W.
   Indoor air quality and HVAC systems / David W. Bearg.
      p.   cm.
   Includes bibliographical references and index.
   ISBN 0-87371-574-8
   1. Heating.   2. Ventilation.   3. Air conditioning.   4. Indoor air pollution.
   5. Air quality management.   I. Title.
   TH7015.B42   1993
   697—dc20   92-29456
                                                                               CIP

COPYRIGHT © 1993 by LEWIS PUBLISHERS
ALL RIGHTS RESERVED

Acknowledgments

There are many people who have helped me to learn as much as I have in the areas of indoor air quality and HVAC systems. First is my father, Milton J. Bearg, P.E., whose experience with HVAC dates back 65 years to when he started with the Cooling and Air Conditioning Corporation in July 1927. His experience as an HVAC design engineer has been very helpful to me over the years.

Other people who have provided guidance, support, or direction for my career, and to whom I wish to express my appreciation include David Gordon, John Spengler, Lou Diberardinis, Steven Larson, and especially William Turner. I have had the pleasure of working with William since I began to specialize in the evaluation of the indoor environment in 1980. Much of the information contained in this book therefore has resulted from my experiences working with William.

In addition to my father, I wish to thank my immediate family, my wife Kate and my two sons Samuel and Nathaniel, who have supported me and put up with me through the long hours required to complete this book.

This book is dedicated to the memory of my mother, Judith Rosenblatt Bearg.

David Bearg is a Registered Professional Engineer living in Concord, Massachusetts with his wife and two sons. He received his B.S. in chemical engineering from Northeastern University, and his M.S. in environmental health sciences from the Harvard School of Public Health. He has been a lecturer at many conferences and seminars and has published numerous papers on air quality, as well as a regular column under the byline of "The IAQ Doctor" for the *Air Conditioning, Heating and Refrigeration News*.

Mr. Bearg has extensive consulting experience through his company, Life Energy Associates, which offers indoor air quality diagnostic and mitigation services, including the assessment of the performance and condition of HVAC systems as well as tracer testing and computer modeling.

Mr. Bearg is a member of the Massachusetts Public Health Association, the National Environmental Health Association, and the American Society of Heating, Refrigeration and Air Conditioning Engineers (ASHRAE). He has been a member of several committees including the Commonwealth of Massachusetts Special Legislative Commission to Investigate and Study the Public Health Effects of Indoor Air Pollution, the National Institute of Building Sciences (NIBS) Committee on Building Envelope Design Guidelines, and the Indoor Air Quality and Ventilation Task team of the Joint Utility Residential New Construction Development project of the Massachusetts Energy Crafted Homes project.

Table of Contents

CHAPTER 1

Introduction

FOREWORD

This book was written to provide a practical guide to those people responsible for the design, installation, operation, and maintenance or evaluation of heating, ventilating, and air conditioning (HVAC) systems. The purpose of this book is to help them understand how these systems are intended to perform with respect to providing and maintaining good indoor air quality (IAQ). The intended audience for this book includes industrial hygienists, HVAC contractors, building owners, managers, facility operators, and those individuals that perform IAQ evaluations. The information in this book is based on experience accumulated over 12 years in the evaluation of over several million square feet of both low- and high-rise commercial (i.e., nonindustrial) office spaces.

OVERVIEW OF BOOK

The organization of this book begins with a detailed description of HVAC systems in Chapters 2 and 3. In Chapter 2, the discussion focuses on the basic types of HVAC systems from a *macro* perspective; that is, as a collection of parts working together. In Chapter 3, the discussion is again on HVAC equipment, this time from the *micro* perspective, where the importance of each of the individual components of typical systems are discussed in terms of the role they play and their importance in terms of the achievement of good IAQ.

Chapter 4 proceeds with a discussion of evaluation criteria for ventilation systems. Discussed first are the specific terms and units that are used for expressing and quantifying ventilation rates. This is followed by a discussion of the standards, regulations, and guidelines that exist for the evaluation of IAQ.

Chapters 5 and 6 present specific discussions of the evaluation techniques available for assessing the performance of ventilation systems. Chapter 5 focuses on how to evaluate the quantity of outdoor air entering the HVAC system, while Chapter 6 discusses how to quantify the amount of outdoor air actually being delivered to the building occupants.

Chapter 7 discusses the characterization of the performance of the ventilation system by discussing the concepts of ventilation effectiveness and efficiency, which refer to the relative ability of the outdoor air being delivered to the occupied areas to dilute and remove air contaminants, as well as to how much of the outdoor air entering the HVAC system(s) actually gets delivered to the occupied areas of the building.

Chapter 8 discusses pressure relationships and the resulting air movement patterns that can exist in buildings. These pressure relationships involve both those between the building and the outdoors and those that exist within the building itself. Included in this discussion is the relationship between these air movement patterns and the achievement of good IAQ, as well as a discussion on how to evaluate these air movement patterns. Examples are given from specific case studies. This chapter also presents a discussion on natural ventilation in buildings.

Chapter 9 discusses the specific tools and techniques available for evaluating ventilation systems, and the information that can be obtained from them. This chapter discusses the investigator as an instrument of evaluation, as well as the use of specific types of equipment.

The concluding portion of the book, Chapter 10, presents a discussion of the potential sources of air contaminants other than those arising merely from the occupants. This last chapter is included to balance out the other side of the dynamic relationship that exists between the amount of ventilation being provided and the sources of air contaminants that are present. Also presented are details on the steps necessary to evaluate the performance of the ventilation systems.

Relationship Between IAQ and HVAC Systems

In order to explain the relationship between IAQ and HVAC systems, it is first necessary to understand how buildings operate and the role that the HVAC system plays in that process. The HVAC system is one of four key elements that interact in a building to yield the conditions of the indoor environment. These four elements are presented in Table 1.1. One primary purpose, or goal, of the building is to provide a healthy and comfortable environment for the occupants of that building. In addition to this concept of the building providing shelter from the elements, there is also the economic reason that buildings are created in order to

Table 1.1. Building Elements Affecting the Indoor Environment

Building shell (its "skin" and partitions)
HVAC system and its condition
Outdoor environment
Building occupants and their activities

bring people together "under one roof" to perform functions that they would not be able to perform as efficiently if they were more dispersed. Buildings therefore are created to increase the productivity of workers. The economic relationship between worker productivity and IAQ is discussed later in this chapter.

The components of the building interact to accomplish these goals. The physical geometry of the space and the allocation of uses within this space bring the people together. The building shell, or envelope, limits the exchange of air and energy between the indoor environment and the outdoors, and the mechanical systems condition the interior air by providing heating and cooling plus making provisions for the exchange of air. The combination of the removal of stale air and the introduction of fresh air dilutes and removes air contaminants, this being accomplished by either mechanical and/or natural ventilation.

Causes of IAQ Problems

An analysis of how a particular building performs these tasks is often essential to understanding the quality of indoor air being provided. One basic unifying premise maintained throughout this book can be summarized in the following statement:

Indoor air quality problems arise in nonindustrial buldings when there is an inadequate quantity of ventilation air being provided for the amount of air contaminants present in that space.

The nature of this dynamic relationship is that, based on the amount of air contaminants present in a particular building, the ventilation rate considered as "adequate" can vary. IAQ problems can also be due to inadequate pollution controls despite otherwise normal or baseline rates of ventilation.

The experience of the National Institute for Occupational Safety and Health (NIOSH)[1] in their IAQ building evaluations provides an example of this dynamic relationship, as evidenced by the breakdown of the causes of IAQ problems as presented in Table 1.2, which summarizes their investigations. Just as the majority of the IAQ problems investigated by NIOSH were attributed to inadequate ventilation, the majority of the information in this book focuses on the requirements for the delivery of adequate ventilation to the occupants of a building. However, since good IAQ is based on both this delivery of adequate ventilation *and* minimizing or eliminating sources of air contaminants, this book also addresses the need to assess the presence of air contaminants.

Table 1.2. NIOSH Indoor Air Quality Investigations by Problem Type (through December 1988)

Problem Type	No. of Buildings	% Attributed
Ventilation inadequate	280	53
Inside contamination	80	15
Outside contamination	53	10
Microbiological contamination	27	5
Building materials contamination	21	4
Unknown	68	13
Totals:	529	100

Defining Adequate Ventilation

Some dictionary definitions for the word "adequate" include: (1) able to satisfy a requirement; suitable; and (2) barely satisfactory or sufficient. With respect to the issue of adequate ventilation, sections of this book include both information on what constitutes adequate amounts of ventilation, as well as how to evaluate whether or not these adequate quantities of ventilation air are being provided. "Ventilation air" is defined as clean, outdoor air delivered to occupied areas of the building which, in conjunction with the air exhausted from the space, dilutes and removes air contaminants present in that space.

Sources of Air Contaminants

In addition to the requirement for adequate ventilation, the other side of the equation for achieving good IAQ is the presence or absence of air contaminants. It is important to remember that: *In assessing the presence of air contaminants, it must be recognized that they can arise from the people who are present in that space, from the activities being performed by these people, from the furnishings that are present in that space, from contamination arising from inadequately maintained HVAC equipment, or from a source located outdoors.*

It should be emphasized that this definition of ventilation is in agreement with the dictionary usage; that is, of providing or introducing fresh air. This interpretation for the word differs from another meaning that has unfortunately come into usage, whereby ventilation has come to refer to the total supply air delivered in order to maintain thermally comfortable conditions. With respect to thermal comfort, it should be noted that no single environment will typically be judged satisfactory by everybody, even if they are wearing identical clothing and performing the same activity. *ASHRAE Standard 55 (Thermal Environmental Conditions for Human Occupancy)* was first published in 1966 and has been revised in 1974, 1981, and 1991. The latest version provides guidance for achieving thermal environmental conditions that "should satisfy at least 90% of the occupants."[2] This means that 10% can be expected to be dissatisfied. It should be noted however that this Standard assumes that people will be dressed the same. In comparison, *ASHRAE Standard 62 (Ventilation for Acceptable Indoor Air Quality)* has as its goal that a substantial majority (80% or more) of the people in a building do not express dissatisfaction upon being surveyed.

The confusion arising from the lack of use of a consistent definition for the ventilation term among all people is aggravated by the practice of using the units of air changes per hour (ACH) for quantifying both ventilation rates and total supply air rates without clarification as to which is being referred to. Whenever the ACH term is used, I recommend that it should also be specified whether it is referring to just the outdoor air portion of the total supply air, or if it is referring to the total quantity of supply air. In this book, the term "ventilation air" will only refer to outdoor air and the total quantity of supply air will be designated as such.

As a guide for those unfamiliar with this distinction, ventilation rates in systems with recirculation typically provide a minimum of 1.0 ACH of outdoor air. The total quantity of supply air can be expected to be in the range of 5.0 to 10.0 ACH. As discussed in more detail in Chapter 4, this amount of supply air typically corresponds to at least 1 cubic feet per minute (cfm) of air per square foot of area and a minimum outdoor ratio of between 14 and 20% of the total supply air.

For systems which do not recirculate air (i.e., 100% outdoor air systems) the ventilation rate and the total supply air rate will be in the range of 5.0 to 15.0 ACH depending on the activity occurring. (e.g., Animal facilities require the 15 ACH.) There are also systems that vary the percentage of outdoor air as a function of the outdoor air temperature. The percentage of outdoor air for these systems, such as those that utilize a constant mixed air temperature approach or those with an economizer cycle, can vary from some minimum value of 10, 15, or 20% up to a maximum of 50 or 100%.

Outdoor Air Quantity vs Supply Air Quantity

The major reason why the volumetric requirements for ventilation air and total supply air are different is because they have two different functions to serve. As stated above, the function of the ventilation air is to dilute and remove air contaminants generated in the space, while the function of total supply air quantity is to achieve and maintain thermal comfort in the space. Since the heating and cooling functions of the HVAC systems are also important in achieving a comfortable working environment, a requirement for good IAQ, the evaluation of the requirements for achieving thermal comfort are also included in this book. The achievement of thermal comfort depends on air temperature, air velocity, mean radiant temperature, and air humidity. Unlike ventilation, which depends on the dilution of contaminated air in the occupied areas with uncontaminated air from outdoors which requires the delivery of an adequate supply of outdoor air to where the people are, the achievement of heating and cooling goals is a function of both the total volume of supply air that can be delivered to where the people are, as well as the temperature of that air. The cooler the supply air, the less supply air that would need to be delivered to a given space to achieve only thermal comfort in cooling mode. However, this situation also presents more of a challenge in preventing drafts for the occupants.

Types of IAQ Problems

The importance of assessing the adequacy of both thermal comfort and ventilation rates can be demonstrated by two typical IAQ problems.

Problem #1. In many older buildings, the original cooling load calculation of internal gains, (i.e., sources of heat in the building) were based only on a certain density of people, lights, and equipment such as electric typewriters. Over time, however, the introduction of computers on every desk has exceeded the cooling capacity of the original HVAC equipment, leading to localized overheating situations and complaints of inadequate IAQ.

Problem #2. For buildings dating from the middle and late 1970s, the HVAC system designs focused on achieving thermal comfort conditions at a minimum energy expenditure, while often ignoring the requirements of adequate quantities of outdoor air for ventilation.

In addition to the premise that IAQ problems result from the delivery of inadequate quantities of clean outdoor air in relation to the amount of air contaminants present, there is also the underlying cause that IAQ problems result from a failure to include IAQ consideration in the decision-making process for the design, installation, construction, operation, and maintenance of the building and its HVAC systems.

IAQ Assessment Approach

Another way of looking at the role of HVAC systems in the achievement of good IAQ is that there are four basic factors that need to be present in order for IAQ to become degraded. These four factors are:

1. A source of air contaminants
2. A person or persons affected by this source
3. A pathway for the transport of the contaminant to the person or persons affected
4. A driving force to transport the contaminant from its source to the person or persons affected

The HVAC system plays a critical role in three of these forces since it not only typically provides the predominant pathway and driving force for air movement in non-residential buildings, but it can also be a source of air contaminants.

In summary, the HVAC system can provide a positive role in the achievement of good IAQ by acting to remove or dilute air contaminants. On the other hand, the HVAC system can also be a major contributing factor in the degradation of IAQ by acting to generate, induce, or distribute air contaminants.

The IAQ evaluation process relies on the scientific method, that is, the creation of a hypothesis followed by the performance of testing and inspections to evaluate the validity of this hypothesis. From the results of this effort, a mitigating control strategy can then be implemented and evaluated. In terms of a specific evaluation protocol for all IAQ building investigations, it must be realized that initially everything involved is suspect and only by eliminating each potential contributing factor, can the precise cause of the problem be identified. It should also be realized that while many buildings are similar, each building is unique, with its own equipment, installation and maintenance history. These steps involved in conducting an IAQ investigation are presented in Figure 1.1. This figure originally appeared in the EPA document, *Building Air Quality*.[3]

Roles of HVAC Systems in IAQ

Specific examples of these actions on the part of the HVAC system can be summarized as follows.

Removal

Examples of the removal of air contaminants include the reduction of particulate concentrations in both the outdoor air and return air streams by the action of the filter bank. Where there is an identified localized source of air contaminants, a mechanical ventilation system can provide control at this source where there are provisions for both the capture and removal of these air contaminants and the delivery of conditioned replacement air to the occupied space. In addition to the removal of particulates, there is also the removal of gases and vapors by such materials as activated carbon, potassium permanganate, silica gel, or alumina. There are new materials being tested and evaluated for this function, and they may well prove to be effective; however, the use of these materials may prove to be too costly to justify in most buildings. In addition, the presence of gases and vapors would in the long run be better dealt with by reducing their potential emissions by eliminating or minimizing their use.

The discussion of the removal of air contaminants by the HVAC systems can also distinguish between the HVAC system providing a local exhaust component, which captures and removes the air contaminants at their source, and the HVAC system performing air cleaning on the recirculated air.

Dilution

Dilution is the most common form of IAQ control provided by HVAC systems in non-industrial buildings. The concept of the "solution to pollution is dilution"

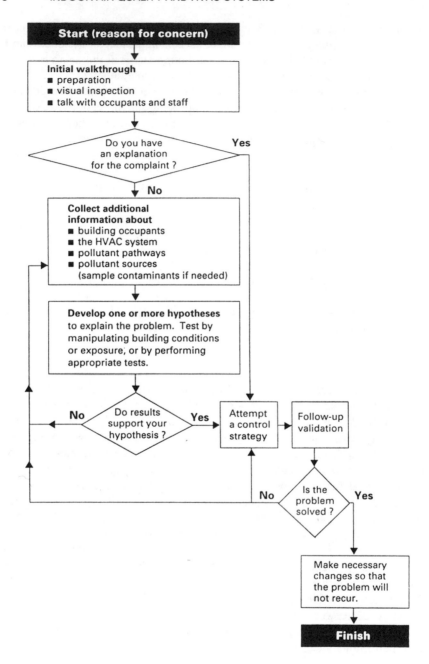

Figure 1.1 Conducting an IAQ Investigation.

has long been recognized as a basic approach for HVAC systems (e.g., for the control of body odor). The first key factor involved is the quantity of clean air introduced into the space being controlled; that is, what is the ventilation rate that is being achieved? This issue is discussed in detail in Chapter 5. The second key factor involved is how does this clean air move through the space being controlled; that is, what are the ventilation characteristics of this system? The variables to be considered are the degree of mixing achieved plus the presence or absence of short-circuiting and stagnant zones, these are discussed in Chapter 6.

Generation

HVAC equipment has the potential to generate air contaminants. This is not an intended event, but it will occur due to deficiencies in the design, installation, operation, or maintenance of the system. One example would be a fan motor with bad bearings that overheats and generates odors that are detectable in the occupied spaces. Or, instead of the bearings, it could be a fan belt that is rubbing excessively and generating particulates that are delivered to the occupied areas. There is also the possibility that during some maintenance activity (such as changing the filters) that if the fans remained in operation, there could be a concentrated release of particulates into the supply ductwork due to poor handling of the filter media.

There is also the special category of generation of air contaminants that relates to the amplification or propagation of microorganisms. In this situation, when the conditions favorable to the growth of biological organisms are brought together, the HVAC system becomes part of the problem. It is important to recognize that the equipment never creates this biological contamination by itself. Rather, it only acts as a host, as an incubator or propagator, or can transport contaminants, but it cannot produce the initial inoculum. This initial inoculum of spores or bacteria is provided instead by either the occupants or from outdoor air, or occurred during the storage of system components.

Induction

By drawing contaminated air into a building, an HVAC can induce, or introduce, pollutants into the occupied areas of the building. This can occur when there is a concentrated source located near the system's air intake. Typical sources include loading docks and building exhausts. It should be noted that unintentional air intakes, such as penetration into the return plenum or the mechanical room itself, can be directly involved in this type of situation. This condition is discussed in greater detail in Chapter 10.

Distribution

Since most HVAC systems typically recirculate a large portion of the air they remove from the occupied space in a building (i.e., return air) back to the occupied spaces, they can redistribute air contaminants from one location to another within the building. Examples can include situations where localized concentrations of air contaminants from activities such as smoking or copying machines are redistributed to locations where these activities are not otherwise occurring. Another example is the situation where differing uses share a common HVAC system, such as in a commercial building where odors from a hair salon are redistributed to the offices of an insurance agency.

PRODUCTIVITY AND INDOOR AIR QUALITY

The relationship between IAQ and the productivity of workers has only recently begun to be investigated. The initial indications are that attempts to save energy in the operation of HVAC systems will be a false economy if these changes lead to decreases in the quality of the indoor environment because of more costly impacts on worker productivity. One such study by Zyla-Wisensale and Stolwijk[4] concluded that, "The cost of providing an optimum workplace environment is expected to be very small compared with the benefit of a productivity increase." This study also found that for the office building investigated, "proximity to supply and return vents and fluorescent lights was related to worker output." This investigation measured productivity by the magnitude of the daily increase in a corporation's data entry file size.

Another measure of productivity for assessing the impact of IAQ is to estimate the cost of absenteeism, or even just the time spent by workers complaining. A feel for the magnitude of this relationship can be obtained by calculating and comparing the cost to an employer of "employee-cost-per-day" with the potential savings from reducing HVAC costs. One such approach by Woods[5] compares these costs on a square-foot basis. Using similar data, the cost values presented in Table 1.3 can be used for discussion purposes.

Table 1.3.   Annualized Cost for a 400-ft² Office on a Per-Person Basis

Cost Component	Total Per Year	Per Person
Total salaries	$90,000	$30,000 (avg.)
Rent ($24/ft²)	9,600	3,200
Amortization of capital equipment	9,000	3,000
Operating costs	2,100	700
Maintenance costs	2,100	700
Energy costs	900	300
Total costs:	$113,700	$37,900

Note: This scenario assumes one manager and two secretaries, 250 working days per year, 2000 working hours per year; capital equipment = furniture, furnishings, and computers, etc., amortized over 3 years.

In this example case, the cost per average employee is $37,900/year, $151.60/day, or $18.95/hour. The value of this employee to the company should exceed this cost; otherwise, there would be little incentive for the company to continue providing employment to this individual. Of this breakdown, the energy costs represent 0.8% of the total. If one were to reduce energy costs by 10% due to reducing the outdoor air quantity, this would represent a potential savings of $90/year. However, if this lead to just one extra lost day by an employee, which is valued at $150 in this example, this would be poor investment.

This relationship can also be considered on a cost per square foot basis. This results in annual costs for salaries, benefit and other support costs in the range of $ 300.00 to $ 500.00 per square foot. This can be contrasted with an annual cost of about only $ 0.50 per square foot to condition 0.2 cfm of outdoor air for ventilation.

The purpose of this example is to stress the importance of achieving good IAQ, not only from the standpoint of employee morale, but also for sound economic reasons. This book is intended to be a useful tool in directing evaluations to determine whether or not HVAC systems are being operated and maintained with this goal in mind.

REFERENCES

1. NIOSH (National Institute for Occupational Safety and Health). Congressional testimony of J. Donald Millar, M.D., Director, before the Subcommittee on Superfund, Ocean and Water Protection, Committee on Environment and Public Works, U.S. Senate, May 26, 1989.
2. *ASHRAE Standard 55*. 1981. American Society of Heating, Refrigerating and Air-Conditioning Engineers, Atlanta, GA. Proposed revisions, 1991, as mentioned by James E. Woods, P.E., Ph.D., in the *ASHRAE Journal,* February 1992, p. 51.
3. U.S. EPA, 1991. "Building Air Quality: A Guide for Building Owners and Facility Managers." EPA/400/1-91/033, DHHS (NIOSH) Publication No. 91-114. S/N 055-000-00390-4. Available from Superintendent of Documents, P.O. Box 371954, Pittsburgh, PA 15250-7954.
4. Zyla-Wisensale, N. H. and J. A. J. Stolwijk, 1990. "Indoor Air Quality as a determinant of Office Productivity. *INDOOR AIR '90. 5th International Conference on Indoor Air Quality and Climate.* Toronto, Canada. pp. 249–254.
5. Woods, J. E. Cost Avoidance and Productivity in Owning and Operating Buildings. *Occupational Medicine: State of the Art Reviews.* Vol. 4, No. 4, October-December 1989. Philadelphia, Hanley & Belfus, Inc. pp. 753–770.

CHAPTER 2

Description of HVAC Systems

OVERVIEW

Before the performance of HVAC (heating, ventilation, and air conditioning) systems can be evaluated, it is necessary to first understand both the specifics of the design and the intended functioning of these systems. This chapter presents a discussion of the identifying basic types of HVAC systems. The discussion includes HVAC system types which provide the normal heating, ventilating, and air conditioning functions in a building as well as systems which provide special ventilation functions such as for smoking lounges and during construction activities. The information presented in this chapter deals with these systems from a *macro* perspective; that is, describing the overall system characteristics. In contrast, Chapter 3 discusses HVAC systems from a *micro* perspective; that is, the role and function of the individual components. This discussion of the individual component parts of HVAC systems includes an explanation of the relationship between each of these components and the achievement of good indoor air quality (IAQ).

FUNCTIONS OF HVAC SYSTEMS

As the acronym implies, HVAC systems are intended to provide for the heating, ventilating, and air conditioning of occupied spaces in buildings. A typical system consists of controls, a minimum of one fan to move the air, a

13

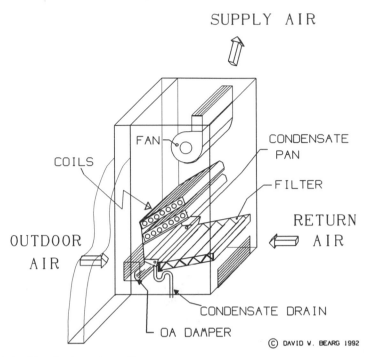

SUPPLY AIR

FAN

COILS

CONDENSATE PAN

FILTER

RETURN AIR

OUTDOOR AIR

CONDENSATE DRAIN

OA DAMPER

© DAVID V. BEARG 1992

Figure 2.1   Sketch of a unit ventilator.

provision for introducing outdoor air, a filter medium to clean the air, coils for heating and cooling the air, dampers for regulating the flow of air, and a distribution system. The distribution system typically consists of ductwork connected to supply registers and a pathway for air leaving the occupied space to return to the air handling unit (AHU).

In its simplest form, this is the basic description of what is commonly called a "unit ventilator." The components of a unit ventilator are depicted in Figure 2.1. In this figure, notice the outdoor air (OA) damper, the filter, the coil, the drain pan, and the fan. These are the basic components of an HVAC system. The importance of each of these individual parts of the whole for achieving and maintaining good IAQ are discussed in Chapter 3.

As the area of the building that the HVAC system serves gets larger, the complexity of the system will typically increase as well. The system may now include such refinements as a more defined mixing chamber where this fresh air is mixed with air returned from the occupied space, distribution ductwork, plus possibly a return air fan. There can also be a provision for a portion of the return air to be exhausted directly to the outdoors, as shown in Figure 2.2 which presents an isometric view of a rooftop AHU. This rooftop system however, displays an all too frequent design deficiency of HVAC equipment. That is, the outdoor air intake is adjacent to the location of the exhaust. This promotes the unnecessary reentry of exhaust air back into the building. It would be a better design if the

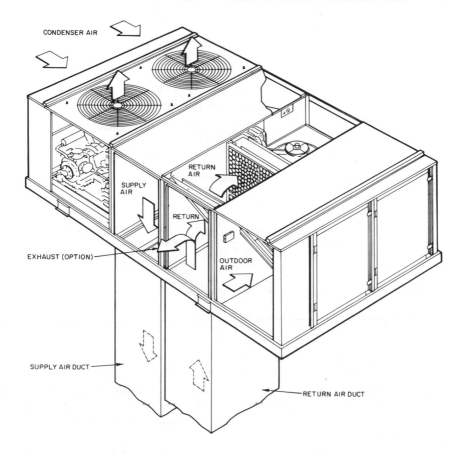

Figure 2.2   Isometric view of a rooftop air handling unit.

outdoor air intake was at least located on the side around the corner away from where the exhaust leaves the unit.

With respect to the exhaust air, it should also be noted that more and more equipment is now being designed to provide heat recovery. This permits the capturing of some of the energy associated with thermal conditioning, either heating or cooling, and then use it to condition the incoming outdoor air.

This provision for air to be exhausted from the building is important because of its controlling influence on the total quantity of outdoor air that can be drawn into the building. After all, the quantity of outdoor air drawn into the building will, over time, be exactly equal to the quantity of air exhausted from the building. This net flow of air through the building is facilitated by either exhaust fans in the building (such as toilet exhausts and building relief vents) or exhaust at the AHU itself. These pathways of air movement for a ventilation system and the building itself are presented schematically in Figure 2.3 for a generic system.

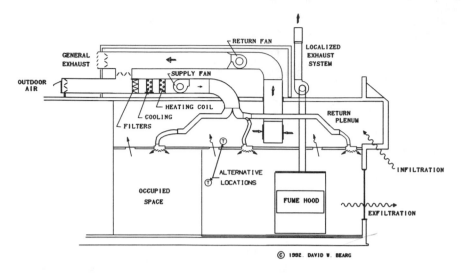

© 1992. DAVID W. BEARG

Figure 2.3 Schematic of a generic ventilation system.

Ventilation

For buildings without operable windows, ventilation is accomplished by mechanical means. A mechanical ventilation system will need to draw in outdoor air, deliver it to where the people are in the building, and then remove stale air to make room for more fresh air. As part of this mechanical ventilation system, there will need to be openings for both the introduction of outdoor air and the elimination of stale air. The mere existence of these openings is not sufficient for this air movement to occur by itself. There also needs to be a driving force — a pressure differential — to achieve the desired air movement; that is, the air will only move as a result of pressure differences. These pressure differences are created both by the operation of the fans and by the action of wind or temperature differences between the indoors and the outdoors.

One important fact to remember is that the quantity of outdoor air entering the HVAC equipment will be a function of both the net open area of the OA dampers and the negative pressure with respect to the outdoors that exists in the mixing box. There are buildings where the OA dampers are wide open, but this "air intake" is in fact functioning as a building exhaust due to a pressure imbalance in the building and the HVAC system, especially the relationship between the supply air and return air fans.

There are other buildings where, although the OA dampers are in their "closed" position, the combination of the leakiness of these dampers and the magnitude of pressure drop across these dampers is so great that the supply air can consist of 30 to 40% outdoor air. The problem with this situation is that the operators no longer can control their building and there can be difficulties relating to the freeze-up of coils, difficultly in maintaining thermal comfort, as well as the potential for wasted energy.

Therefore, just noting the position of the OA dampers is an insufficient determination for knowing if outdoor air is coming in because the OA dampers could be "closed" but sufficiently leaky and there be a sufficient pressure drop across them that adequate quantities of outdoor air are being introduced, while it could also occur that the OA dampers are "open" but there is also a pressure imbalance occurring so that no outdoor air is being introduced.

Building Pressurization

Included in the proper functioning of HVAC systems, therefore, is the requirement of maintaining these systems and the building at appropriate pressures with respect to the outdoors. The occupied areas of the building should be maintained at a slight positive pressure, typically 0.05 in. of water column with respect to the outdoors in order to minimize the infiltration of unconditioned air into these areas of the building. Other units to express this relationship would be 12.5 Pa, 0.26 psf, or 0.0018 psi. This pressurization of the occupied spaces is achieved by operating the supply fan, or dedicated OA fan, so it attempts to introduce a larger volume of air into the building than the other AHUs are attempting to remove from the building. Since the quantity of air that can be introduced into the building must equal the quantity leaving, a net overpressure is achieved in the occupied areas of the building. It should be noted that other portions of the building envelope, such as where the return plenum is adjacent to the outdoors, will therefore be under a negative pressure with respect to the outdoors.

In addition to the mechanically driven (i.e., by fans) pressure differentials across the building shell, there are also the pressure differentials created by the action of the wind and by temperature differences between the indoors and outdoors. The significance of these thermally driven pressure differentials is that for buildings that are heated, air will escape at upper-level penetrations. This loss of air from the building then has the potential to create negative pressure in the lower level of the building which will draw in outdoor air at lower level penetrations. The unintentional introduction of outdoor air into the building at lower levels can be aggravated by the location selected for the mechanical room. This is because the mechanical room will typically be under negative pressure with respect to the outdoors due to unsealed metal-to-metal connections and penetrations in the AHUs. This also means that air contaminants arising from sources located either in the mechanical room or nearby (such as loading docks) can be drawn into the HVAC system and distributed throughout the building. The evaluation techniques useful for characterizing this situation and other pressure relationships are discussed in detail in Chapter 8.

OVERVIEW OF TYPES OF HVAC SYSTEMS

One of the first basic distinguishing characteristics among different HVAC systems is whether the system is *decentralized* around the building into many similar AHUs, or whether there is a *centralized* AHU or AHUs. Buildings, of course, can utilize a combination of these two approaches, with unit ventilators dispersed around the perimeter of the building and the core areas served by a centralized system. Decentralized systems are frequently selected for their lower first costs, which are achieved by eliminating the cost of ductwork. Over the life of the building, however, this can be a false economy, as they can become difficult and time consuming to maintain. It should be remembered however that even HVAC systems that are primarily centralized, with respect to the location of the AHUs, will also have numerous decentralized components that will require periodic inspections and maintenance. Examples of this situation are systems with distributed reheat coils in the supply ductwork or perimeter fan-coil units.

The next major distinction among HVAC systems to be aware of when performing an IAQ evaluation is the basic approach employed for achieving thermal comfort. There are two basic techniques: constant air volume (CAV) systems and variable air volume (VAV) systems. In CAV systems, thermal control is achieved by delivering a constant volume of air to all locations and varying the temperature of the delivered air to each location. In contrast, the VAV systems utilize a resetable constant temperature of the delivered air to most locations, while varying the quantity of air delivered to the individual zones in the building.

Another distinction among the types of HVAC systems refers to the heat transfer mediums used: (1) all-air systems, (2) air-and-water systems, and (3) all-water systems. All-air systems cool the occupied spaces in buildings by delivering cold air to these areas. Heating can either be accomplished by the same air stream or by a separate heating system. If the heating is accomplished by the air stream, it can be added either in the central system or at a particular zone (terminal reheat system). Air-and-water systems condition spaces by distributing air and water sources to terminal units installed in habitable space throughout a building. The air and water are cooled or heated in central mechanical equipment rooms. All-water systems heat and/or cool a space by direct heat transfer between water and circulating air and typically have no provision for providing ventilation.

Another distinction among different types of HVAC systems is that some equipment is set up to handle just one zone, whereas others can handle multiple zones. In evaluating an HVAC system, it is important to know what constitutes the boundary between zones. The separation of differing zones can be centralized or decentralized. That is, for example, a centralized approach for CAV systems will have the appropriate mix of warm and cool air enter ductwork at the AHU that only goes to a particular zone. In the decentralized approach, the ductwork delivers both the warm and cool air to a mixing box near the zone and the apportionment occurs at this location.

Discussions of these basic differences in HVAC system design as they relate to the achievement of good IAQ occupy the balance of this chapter, which is organized into the following categories:

- Decentralized HVAC systems
- Centralized HVAC systems:
    Constant air volume (CAV) systems
    Variable air volume (VAV) systems
    Hybrid HVAC Systems
    Heat pump HVAC systems

Decentralized HVAC Systems

A typical decentralized HVAC system consists of individual unit ventilators dispersed around the perimeter of the building (see Figure 2.1). The decision to go with this type of system typically was based on their relatively low first costs and degree of control, and are frequently found in places like schools. These unit ventilators include, in addition to fan-coils for heating and sometimes cooling, a penetration through the wall with a damper to permit the introduction of outdoor air. The downside of this approach is that with large numbers of unit ventilators distributed throughout the building, they represent a large commitment in terms of maintenance.

Another challenge with the use of unit ventilators is preventing freeze-ups of the coil while still maintaining the introduction of adequate quantities of outdoor air and air economizer capability. In some buildings where they have experienced freeze-ups of coils, the outdoor air intakes for all the units have been permanently closed to prevent future occurrences of this problem. In some buildings, the dampers are just closed off during the coldest winter months or have been permanently disconnected as part of an effort to reduce energy costs.

Owing to the labor-intensive requirements of performing maintenance on these units, records of reported maintenance activities may not be accurate. In one building, it was reported that a unit ventilator had been serviced 6 months ago. Upon inspection, however, it was obvious that the unit had not been opened since the unit had last been painted. The records of when the painting had been performed indicated that this unit had been painted 2 years ago. This is an example of where the IAQ investigator is the recipient of second-hand reports of the condition or operating parameters of HVAC equipment that are not supported by the physical evidence.

Another example of this situation was when I was shown work orders for the task of reattaching linkage arms controlling the position of outdoor air louvers for the unit ventilators in a school building. This work was to be performed to correct what had been an overzealous attempt at energy conservation. However, upon inspection, it was determined that many of these linkages were still disconnected.

I have therefore found it necessary to be very skeptical when listening to descriptions of work that has been accomplished when the source is other than the person who did the work.

Thus, when evaluating the IAQ in a building which receives some or all of its ventilation air from perimeter fan-coil units, or unit ventilators, an inspection should be performed of a representative number or all units to assess the functional condition of all components of the outdoor air path. At a minimum, this should include the following.

1. Checking that the outdoor air grille is not obstructed
2. Checking that the linkages that operate the OA damper are present and functional
3. Quantification of the amount of outdoor air being introduced by each unit
4. Appropriate response of thermostat

The third item can be achieved by comparison measurements of carbon dioxide (CO_2) or sulfur hexafluoride (SF_6) concentrations in the return air and supply air streams. This procedure is discussed in detail in Chapter 5.

Centralized HVAC Systems

The majority of commercial (i.e., non-industrial) buildings will typically include a centralized air handling system. A centralized system will require the allocation of space from the building for the location of the HVAC equipment, although it will usually be located so as to minimize the consumption of rentable space. There is a tremendous amount of variation in the locations selected for the HVAC equipment; sometimes it is in the basement, sometimes in a penthouse mechanical room, sometimes there are both, sometimes it consists of packaged rooftop units. The size of the building and the corresponding size of the equipment will of course play a major role as to where this equipment ends up being located. Unfortunately, the emphasis on maximizing profit and the quantity of rentable space results in pressures and decisions which minimize the space available for the mechanical equipment. All too often, this equipment is relegated to locations which are not easily accessible. The potential problem with this approach is summarized in an axiom of mine: "The more difficult it is to perform maintenance on HVAC equipment, the less likely this maintenance will be accomplished." Too many HVAC systems are designed without sufficient recognition of the fact that this equipment will need to be serviced and maintained over its useful lifetime.

HVAC equipment can be comprised of a self-contained package unit, or it can consist of individual components that were assembled together at the site. This latter type is referred to as a built-up system. Built-up systems are typically located in mechanical rooms which can be in a rooftop (either in a penthouse enclosure or out in the open), in a dedicated mid-floor location, in the basement, or divided among a combination of these locations. The distinction between package units and built-up systems are that package units are fabricated in a factory and delivered to the building with their major components (including controls) already assembled and wired. The installation of these units is therefore limited to placement of the unit plus connection to controls, power supplies, and the ductwork for return air, supply air, and, where necessary, outdoor air ductwork.

Comparison of CAV and VAV Type Systems

As compared with the decentralized perimeter unit ventilators which are constant air volume systems, centralized systems are either CAV or VAV systems. The basic distinction between these two types of systems is how they attempt to achieve thermal comfort in the spaces they serve; however, this difference leads to very important differences in terms of IAQ.

CAV systems are designed to respond to variations in thermal loads among different locations by varying the temperature of the air delivered to a given zone in the building, while VAV systems attempt to respond to the differing thermal requirements by varying the quantity of air delivered to that zone. These two approaches to achieving thermally comfortable conditions in the building have significant potential impacts on the amount of ventilation air being provided in those buildings.

Constant Air Volume (CAV) Systems

As implied by it name, CAV HVAC systems deliver a constant volume of supply air. In order to maintain thermally comfortable conditions within different zones in the building, this approach varies the temperature of the air delivered to the occupied spaces. There are three ways to accomplish this: (1) there is a single zone, the delivery temperature is set as it leaves the AHU, and the distribution is based on a single-duct system; (2) there are multiple zones and each zone has ductwork that separates at the AHU; or (3) there are multiple zones and the temperature of the delivered air for each zone is adjusted at a mixing box serving that zone.

In the last approach, the distribution system consists of two sets of ductwork, one transporting heated air, while the second has cooled air plus mixing boxes. A schematic of a dual-duct CAV system is presented in Figure 2.4.

The coils that thermally condition this air are referred to as the hot deck and the cold deck, respectively. In a properly controlled system, the hot deck only heats the air to be able to provide thermal comfort conditions in the coldest zone of the building, while the cold deck only cools the air to a temperature sufficient to achieve thermal comfort conditions in the warmest zone of the building. Zones in the building with intermediate needs for heating or cooling receive a combination of the two air streams, which are apportioned by a mixing

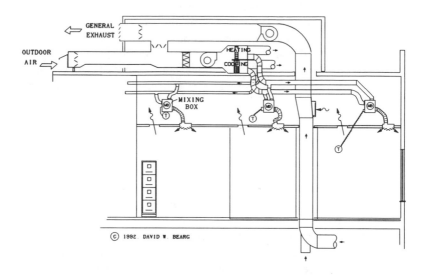

Figure 2.4 Dual-duct constant air volume HVAC system.

box or terminal unit. In the mixing box is a valving device which varies the proportion of the quantities of warm and cold air delivered to the space in response to the thermostat serving that particular zone. One of the distinguishing characteristics of a CAV system is that if the location being served requires neither heating nor cooling, the delivery temperature then becomes the same as the in space temperature but the total volume of air continues the same as if full cooling or heating were required.

In a dual duct constant volume system, the mixing box only controls the temperature of the delivered air. However, in larger systems, the mixing terminal unit may maintain both the temperature and the air volume. This then becomes a variable volume dual duct system which will also include a system of static pressure regulators in the ducts.

Variable Air Volume (VAV) Systems

As the name implies, VAV HVAC systems vary the volume of air delivered by the HVAC system. In order to maintain thermally comfortable conditions within different zones in the building, this approach varies the volume of air delivered to each zone, while the air temperature to most locations remains constant. Some locations, typically perimeter areas or areas where dehumidification is desired, have terminal reheat boxes in the distribution system so that the delivered air can be a higher temperature in order to compensate for heat loss at the periphery of the building. In a commonly designed basic VAV system,

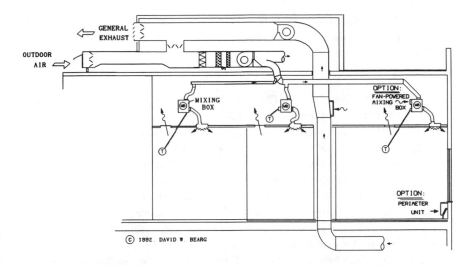

Figure 2.5   Schematic of variable air volume system.

shown in Figure 2.5, the total air flow rate is modulated in response to the changing building heating or cooling requirements.

The maximum total air flow is designed to be able to maintain thermally comfortable conditions for the more stringent of either winter or summer conditions. Winter design conditions are based on temperatures that have been equaled or exceeded 99 or 97.5% of the time during the winter months. Summer design conditions are based on dry-bulb temperatures that have been equaled or exceeded 1, 2.5, or 5% of the time during the summer months. These design conditions will be reflected in the total air flow rate, the ventilation rate capability, and in the heat transfer capacities of the cooling and heating coils. During the rest of the year during part-load conditions when the thermal loads are reduced, the normal operation of a basic VAV system results in a decrease in the total air flow rate. If the outdoor air control approach merely relies on the quantity of outdoor air introduced being a constant percentage of the total supply air, then with the total volume of supply reduced this can result in the level of mechanical ventilation falling below ASHRAE minimum recommended values. As pointed out by Roberts,[1] "Traditional control systems for air handling units and variable volume terminals do not include provisions for maintaining required outdoor air flows to the various spaces."

Example 2.1: Building Operating at Design Conditions. This is a 100,000-ft^2 building with a VAV system with a design capacity of 100,000 cfm (1 cfm/ft^2) and an outdoor air control approach of maintaining a constant 20% of the supply air as outdoor air. There are 700 people in this building, so that at design conditions, 20% of 100,000 cfm equals 20,000 cfm; thus, there is theoretically 28.6 cfm of outdoor air per person.

Remembering that design conditions will exist for only a small percentage of the time throughout the entire year, what needs to be considered is the amount of ventilation being provided at other than design conditions — the part-load situation.

Example 2.2: VAV Building Operating at Part-Load Conditions. Consider the same 100,000-ft^2 building with its VAV system design capacity of 100,000 cfm (1 cfm/ft^2), an outdoor air control approach of maintaining a constant 20% of the supply air as outdoor air, and an occupancy of 700 people. At a part-load condition resulting in a total air flow of 50,000 cfm of supply air (a not unreasonable situation) 20% of this 50,000 cfm now equals 10,000 cfm, so there would only be a potential for 14.3 cfm of outdoor air per person. *Therefore, a VAV system relying on an outdoor airflow rate that is a constant percentage of the total air flow rate will typically not be able to provide adequate quantities of outdoor air for ventilation under part-load conditions.*

Recognizing the importance of controlling the quantity of outdoor air to ensure that minimum standards are maintained under changing supply air volumes, it is necessary that the minimum outdoor air flow rate be established as a controlled parameter in the design criteria. There are several approaches for achieving this goal. One approach would be to have the minimum outdoor air flow quantity occur at the supply fan's minimum anticipated volume. This approach, however, as pointed out by Gardner,[2] could result in energy waste at other than this minimum flow condition.

Another approach for guaranteeing that the quantity of outdoor air is sufficient for the occupant ventilation requirements is the use of some form of ventilation control, either partial or full. One technique is to control the interlocked OA damper and return air damper in response to an outside air velocity sensor or air flow measuring device. This approach is depicted in Figure 2.6.

A less costly and less closely controlled variation on this concept deletes the OA damper control and modulates only the return air damper in response to the outside air flow rate or velocity sensor. A straight length of ductwork and a reduced duct cross-section may be required to accurately sense the outside air flow rate. In lieu of a velocity sensor, a perforated plate may be installed in the outside air duct with a static pressure sensor that controls the return air damper, thereby maintaining a constant air flow rate.

If the ventilation requirement is greater than the expected minimum supply air flow rate, then a supply air low limit controller should be incorporated into the system to maintain the minimum design outside air flow rate. This configuration is depicted in Figure 2.7.

Another approach for maintaining an adequate quantity of outdoor air with a VAV system under part-load conditions is with a suitably sophisticated DDC (direct digital control) system that can increase the outdoor air percentage of the supply air as the quantity of total supply air decreases, so as to maintain a constant absolute minimum quantity of outdoor air entering the

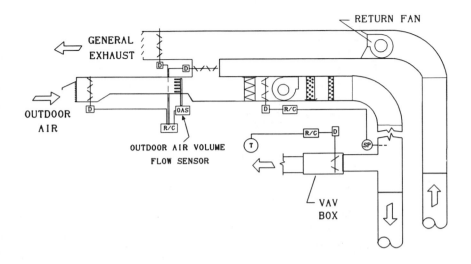

Figure 2.6   VAV system with partial ventilation control.

HVAC equipment. As with the previous technique, there needs to be a sensor for measuring the outside air flow rate.

A different approach for maintaining an adequate quantity of outdoor air with a VAV system under part-load conditions that has been demonstrated to work successfully is with the use of a separate dedicated constant volume outdoor air fan. In this type of installation, the VAV system has full economizer capability and, when the system logic calls for the minimum outdoor air condition, then the economizer OA dampers close and the OA dampers for this separate OA fan open fully and this fan comes on. This arrangement, as depicted in Figure 2.8, is therefore able to provide the necessary outdoor air quantity no matter what the part-load total supply air quantity has been reduced to.

Another approach also uses a dedicated outdoor air fan, but this time there is also a heat recovery unit, such as an air-to-air heat exchanger, which extracts heat, or coolth, from the exhaust air stream in order to temper the incoming air.

VAV Distribution Issues

Another aspect of VAV systems that has the potential to adversely affect IAQ relates, not to the quantity of outdoor air entering the HVAC equipment, but to the localized zones served by individual VAV mixing boxes or diffusers. This microcomponent of the system, like the macro behavior of the system, varies the volume of air delivered to one zone in response to the thermostat of that zone. If that zone has reached thermal equilibrium, then this zone control may close off the

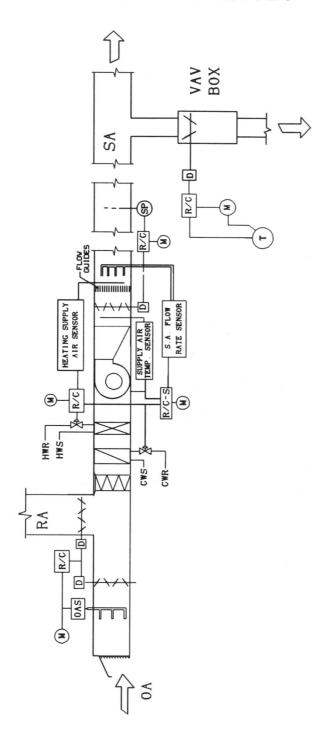

Figure 2.7 VAV system with ventilation control.

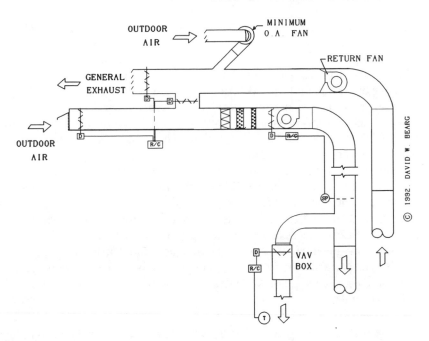

Figure 2.8   VAV system with separate minimum outdoor air fan.

flow of air entirely or it may close down to some preset minimum flow, usually 0 to 50% of the total. In the first situation and possibly in the second as well, the local inhabitants of this zone may not be receiving the adequate outdoor air quantities. If local zones are shutting down or going to their minimum positions, one system modification to consider is to reset the temperature of the delivered air slightly warmer. At these locations, more air will be needed to maintain adequate cooling. The problem with this approach is that other locations may overheat due to the fact that the delivered air is too warm to effectively cool that area. In situations like this, localized areas are overheating typically because of concentrations of heat generating equipment; supplemental cooling can be provided separate fan-coil terminal units. A common brand name for this type of equipment is Liebert™, but the term "Liebert unit" is almost a generic description. *Therefore, when performing an IAQ evaluation of buildings with VAV systems, the investgation should include the measurement of a representative number of diffusers or mixing boxes to quantify the minimum setpoints of this supply air distribution equipment.*

Hybrid HVAC Systems

Some HVAC systems are neither strictly of the CAV or VAV type, but include aspects of both types of systems. For these systems, the evaluation should compare how thermal comfort is achieved in perimeter areas of the building in contrast to the core areas of the building. This distinction exists because the perimeter areas need to be able to respond to a much greater range of temperature variation. In many climates, perimeter areas must overcome freezing outdoor temperatures and localized heat from solar gain. The thermal loads for interior portions of buildings typically vary far less, needing only to provide cooling to balance the heat given off by people, equipment, and in some buildings, the lights. Some equipment such as large photocopiers can generate large amounts of heat. The significance of this approach, in terms of IAQ, is not only the issue of the quantity of outdoor air being provided, but there can also be issues relating to the quality of the supply air delivered.

Another issue is the relationship between this source of heat and the distribution of the supply air. In some buildings, the recognition of large copiers as concentrated sources of heat has led designers to deliver large amounts of supply air directly to their location. The problem with this approach is that since copiers are also significant sources of air contaminants, these contaminants will then be distributed throughout the space. A better approach is to deliver the supply air to the people and place more return registers, or preferably exhausts, over the copier, as close to the source as possible, to remove both the heat and air contaminants that they generate.

Example 2.3: Building with Fan-Powered Mixing Boxes. In one building investigated, there were IAQ problems attributed to the hybrid HVAC system used. The core area, where there was a fairly constant cooling demand, was served by a VAV system with no problems. The perimeter system, however, included fan-powered mixing boxes which could vary both the quantity of air delivered as a function of the operation of the fan and could also vary where this air was coming from; either entirely from the centralized system or it could utilize the warmer air from the adjacent return air plenum. Figure 2.9 displays a drawing of a fan-powered mixing box.

When air was drawn directly from the return air plenum, to provide additional heating to the perimeter areas, this air received only minimal filtration. Although these fan-powered mixing boxes had been modified to include filters, the filter

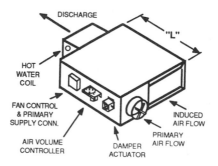

Figure 2.9   Fan-powered mixing box.

media typically was poorly fitted or deformed. Since the return air plenum was laden with dust from construction activities, the use of this perimeter system for supplemental heating resulted in the introduction of high levels of particulate matter in the occupied spaces. This was in strong contrast to the air delivered directly from the centralized HVAC equipment which had high efficiency filters that were well installed and replaced at appropriate intervals.

As part of the investigation to determine the source of the excess particulates in the space, a qualitative assessment of local dust levels was achieved by a finger wipe of the VDT (video display terminal) screens on the desks. The hypothesis had been put forward that the carpets were a major source of particulate matter in the air. However, this crude investigation indicated that dust levels were higher in the perimeter locations than in the core locations. A similar result could also be due to infiltration if there was a pressure imbalance. This result led to a more in-depth look at the perimeter system. *Therefore, whenever a building under evaluation is determined to include fan-powered mixing boxes, the inspection should include a determination of the adequacy of the filtration achieved by these distributed units.*

Heat Pump HVAC Systems

Another category of HVAC systems rely on heat pump units, typically of the water source type. These units are typically located in the ceiling plenum space (above the suspended ceiling) and are distributed around the building. With this approach, thermal comfort is achieved separately for each zone served by each individual heat pump unit. These heat pump installations are typically constant minimum outdoor air setups. Outdoor air is either ducted to either just the plenum space or to the vicinity of each heatpump unit. From the IAQ perspective, the problems that can occur with many of these systems relate to an insufficient quantity of outdoor air reaching the heat pump units. This can be due to an undersizing of the outdoor air supply system or because the outdoor air is merely dumped into the plenum and does not find its way to each of the heat pump units. With the delivery of inadequate quantities of outdoor air, the heat pump units merely recirculate the same air, albeit thermally conditioned. The other IAQ issue for this type of installation is that by distributing many individual heat pump units through the building, a tremendous maintenance challenge is created because of difficulties in physically accessing the equipment and because they are located in the occupied areas of the building as opposed to isolated mechanical rooms. In some buildings, remodeling efforts have in fact made access to some of these units effectively impossible. *Therefore, whenever a building under evaluation is determined to be based on distributed heat pumps, the inspection should include a determination whether the source of outdoor air is ducted to each unit, as well as the quality and installation of the filter, and whether the condensate pan is clean and drains completely.*

CONTROL OF THE HVAC SYSTEM

Just as the complexity of the equipment under the car hood has increased over the years, there has been a similar phenomenon with respect to controls for HVAC systems. This section is just a basic primer on the subject of HVAC controls emphasizing just what is needed to know for performing a ventilation-based IAQ evaluation. As part of the evaluation effort to determine if the HVAC system is performing as intended, it is necessary to consider not only the quantity of outdoor air expected to be delivered, but that the equipment and its controls are performing as intended. When reviewing the schedule of equipment on the mechanical drawings, the sequence of operations should also be examined. The sequence of operations, as its name implies, describes how the system is turned on, shut down, how it is supposed to respond to changes in outdoor air temperatures and defines operation during occupied and unoccupied conditions.

Thermal Control Approaches

As pointed out in the discussion comparing VAV systems with CAV systems, HVAC systems designed to merely achieve thermally comfortable conditions can be deficient in their achievement of adequate amounts of ventilation. With respect to maintaining thermally comfortable conditions, the basic categories of control systems can be summarized as follows.

Self Contained

This control system is powered by the control media. An example is a liquid- or gas-filled capillary bulb which, by means of a bellows, controls the position of a valve for perimeter radiation. The advantages of this approach are that it is simple, inexpensive, and has low maintenance. The disadvantages are that it has a slow response, there is no communication to other systems, and they only maintain a given set point to only about $\pm 2°F$.

Electric

This control system is powered by electricity. An example includes a bimetal temperature-sensing element which makes or breaks an electric circuit controlling an actuator which controls the position of a valve for perimeter radiation. The advantages of this approach are that it is simple and inexpensive. The disadvantage is that they can only maintain control of a given set point to about $\pm 2°F$.

Pneumatic

This control system is powered by compressed air. This approach is widely used with HVAC equipment where variations in the pressure of the compressed air modulate the position of valves and dampers. The advantages of this approach are that it is relatively inexpensive and many people are familiar with its use. The disadvantages are that it requires a source of clean and dry (filtered and cooled) regulated air and an air compressor and it can be prone to controllers going out calibration and catastrophic failure. These failures result when compressed air leaks somewhere in the system. When this occurs, it is typically first discovered when the air compressor begins operating more frequently. Sometimes, this results in the air compressor burning itself out, thus effectively shutting down the control system until both the compressor is repaired or replaced and the leak is found and eliminated. Finding the leak can be difficult to locate among the hundreds or thousands of feet of primary and secondary control lines of copper or plastic tubing, $1/8$ to $1/2$ inch in diameter.

Electronic

This control system uses an electronic circuit to convert a sensor reading into an output signal in the range of 0 to 24 V DC. This analog signal has the advantages of accuracy plus an ease and stability of calibration. The disadvantages are that it is relatively expensive and susceptible to electrical noise.

Direct Digital Control (DDC)

This control system relies on logic in a computer program to convert digital and analog sensor inputs into digital or analog outputs to control the operation of the HVAC equipment. The advantages of this approach include its flexibility in control strategies, accuracy, ease of operation with graphical displays and control settings, monitoring and record-keeping abilities, and reduced maintenance. Its disadvantages include the need for more sophisticated training on behalf of the operators. In this approach, the sequence of operations is converted to software in the form of algorithms. Sometimes, the documentation of this computer program is not maintained and can lead to a situation where the building is operating on its own in accordance with the instructions in this code while the "operators" have lost control of the system.

Example 2.4. In one building where a DDC system was installed as part of an upgrade to the HVAC system, a disagreement between the HVAC contractor and the building's owners led to a parting of the ways. Unable to resolve their differences, the HVAC contractor was fired and did not receive his invoiced full payment. They in turn failed to provide a copy of the software controlling the building's system, leaving the building to operate on its own.

A potential problem for all of the controls is that they are only as good as their sensors; calibration is critical.

Example 2.5. In an investigation of a 12,000-ft^2 building with two floors, located in a northern state, the four packaged AHUs, each rated at 3050 cfm, were all controlled by a single mixed air sensor/controller. This sensor appeared to be seeking a mixed temperature of 60°F, based on the position of its dial. In fact, many people who checked this building assumed that it was set correctly. However, upon placing a remote reading thermometer in the mixing chamber of one of the AHUs, it was determined that this sensor/controller was actually seeking 70°F. The consequence of the one controller being out of calibration, whether intentional or not, was that it kept the OA dampers closed on all four of the AHUs, thus depriving all of the occupants of that building of an adequate supply of outdoor air for ventilation.

Example 2.6. In an investigation of a building at a college, where the number of buildings had increased significantly while the number of maintenance personnel had not, the first thing noticed upon entering the mechanical room on a very hot day was the banging of the ductwork every 7 seconds as both the OA and return air (RA) dampers alternated between being in their full open and then the fully closed positions. It was the RA ductwork in the mechanical room which was banging as it alternately was under suction, when the OA dampers were closed and the RA dampers were open, and then pressurized, as the RA dampers closed and the OA dampers opened. This malfunction was traced to a string of events. First, the make-up float valve for the chilled water supply had failed and was on order to be replaced. In the absence of this device, make-up chilled water needed to be added manually and the system was low on water. As the outdoor air temperature warmed up to above that of the return air temperature, the control logic for switching from full economizer to minimum outdoor air, as would be expected under these conditions, depended on two control functions: one was the temperature of the chilled water and the other was a signal from a flow measuring device that indicated that the chilled water was flowing. The temperature signal indicated that the system should reset to minimum outdoor air; however, the flow sensor could only provide an intermittent signal on chilled water flow due to the lack of a sufficient quantity of water in the system. This example indicates that there are no unimportant components in a HVAC system. This situation reminds me of the epic poem that goes, "For want of a nail, a shoe was lost. For want of a shoe, a horse was lost. For want of a horse, a rider was lost, etc." More information on the specific importance of individual parts in an HVAC system is discussed in Chapter 3.

Example 2.7. In an investigation of a multistory building with many AHUs and several mechanical rooms, temperature testing of the OA, RA, and mixed air (MA) streams was being performed to determine the percentage of outdoor air entering the AHUs in their minimum outdoor air condition. The central computer had instructed all AHUs to have their OA dampers in their minimum position for the purposes of this testing. Upon entering the mixing chamber of one AHU, the coldness of the space immediately indicated that the OA dampers were not closed.

Upon additional investigation, it was determined that there was a local switch which had disabled this AHU from control by the central computer, unbeknownst to this device or the operators. Once again, there needs to be a certain degree of skepticism on the part of a good IAQ investigator to be able to prove that systems and their controls are functioning as intended and that this intent is appropriate for the current use of the building.

Controls Affecting the Quantity of Outdoor Air

Decisions are made during the design of a building which affect the ability of the HVAC equipment to be able to introduce and deliver adequate quantities of outdoor air. Some building designs have placed an over-emphasis on energy conservation over IAQ by selecting systems which can only provide *a constant minimum quantity of outdoor air*. The only advantages to these systems are the simplicity of their controls and the ease with which they can be evaluated with respect to their ventilation rates. That is, other systems for controlling the quantity of outdoor air are more sophisticated in that the quantity of outdoor air can vary as a function of the outdoor air temperature. The fact that the quantity of outdoor air drawn in by many HVAC systems can vary during the day, if the outdoor air temperature changes, raises an important issue for evaluating the performance of HVAC systems.

IAQ and energy conservation do not have to be an "either/or" proposition since air economizers can save energy as compared with the fixed minimum outdoor air approach in many climates. The trade-off with this approach in some building configurations is the requirement for larger ducts needed to distribute the outdoor air to the individual floors. IAQ and energy conservation can also be attained by the use of the control approach called "demand-controlled ventilation," where the amount of ventilation provided is a function of the actual number of people present in the zone being controlled.

Whatever approach is being used, however, the following recommendation is crucial: *before the performance characteristics of an HVAC system can be evaluated, the investigator must be able to document the operating conditions of that system during the time that the testing is being performed.*

One basic control approach for regulating the amount of outdoor air entering the HVAC relies on achieving a *constant mixed air temperature*. With this approach the quantity of outdoor air is varied so that the combination of the RA stream and the OA stream yield a MA stream at the desired delivery air temperature. Depending on the outdoor air temperature, this control regime varies the position of the OA damper or dampers from a minimum position up to a maximum position, with the minimum outdoor air position occurring at the temperature extremes of very warm and very cold outside air. At the high temperatures (i.e., above the desired delivery temperature or above the return air temperature depending on the individual system), the OA dampers are reset to their minimum outdoor air position.

Demand Controlled Ventilation

Demand-controlled ventilation (DCV) is an approach for regulating the quantity of outdoor air delivered to a zone in response to the number of people present in that space. Just as occupancy sensors are being used to save energy by turning off lights when spaces are unoccupied, DCV can save energy in certain situations of low occupancy. In order for this approach to be cost effective, the occupancy needs to be variable, such as would be occurring in a theater, an auditorium, a meeting room, or a classroom.

The way that DCV works is that some component of the indoor air, such as carbon dioxide (CO_2), is monitored as an indicator of the number of people in the building, and the measurement of this concentration is input data to logic circuitry that can adjust the position of the OA dampers. This approach has the potential to both help in the achievement of good IAQ and conserve energy by making sure that as much outdoor air as can be delivered by the system is available when the space is fully occupied, but that overventilation is not occurring when occupancies are low or zero.

This approach capitalizes on the fact that just as the increase of CO_2 concentrations indoors over outdoors can be used to assess the adequacy of the amount of ventilation being provided in a space, a controller that could measure CO_2 concentrations can be used to optimize the operation of ventilation equipment. Recognizing that the number of people in a building can vary considerably throughout the day, relying on just a constant minimum amount of ventilation during the occupied portion of the day can result in a ventilation rate that can be either greater than required or less than required. Using a ventilation control strategy that is more directly based on the number of people that are actually present in a given space has the potential to create a more optimal solution, where both energy savings and adequate or good IAQ can be achieved.

To achieve this control strategy, there needs to be a CO_2 analyzer/controller which can incrementally open the OA damper when the CO_2 setpoint or setpoints, (e.g., 600 to 800 ppm) is approached. Conversely, there would also need to be another lower CO_2 setpoint or setpoints, (e.g., 750 ppm) which would indicate when the OA dampers would incrementally close one stop.

If this approach were in use and the CO_2 concentration in the space never exceeded 800 ppm, the resulting ventilation rate would depend on the outdoor air concentration. Assuming the outdoor value was 350 ppm, then an average of 24 cfm of outdoor air is being provided for each person in the building. Similarly, if the not-to-exceed value were 700, 900, or 1000 ppm and assuming the same 350-ppm concentration outdoors, then not exceeding 700 ppm indoors corresponds to 31 cfm per person of outdoor air being delivered on the average. By the same calculation (see Equation 6.2), 900 ppm corresponds to averages of 19 cfm per person, and 1000 ppm corresponds to 16 cfm per person.

In terms of measuring the CO_2 concentration, the measurement location for the CO_2 concentration should be representative of the space being controlled.

Therefore, one approach is to monitor the CO_2 concentration in the return air duct. However, one should first verify that the CO_2 concentration measured in this return air duct is representative of the space being controlled. As discussed in Chapter 6, the measurement in the return air duct may not be representative of the space this air is coming from because if the supply air distribution system is leaky and is permitting supply air to short-circuit directly into the return plenum without actually reaching the people, then the CO_2 concentration in the return duct will be lower than in the space, having been diluted by the lower CO_2 value of the supply air. This means that by performing a comparison of the measurements of CO_2 concentrations at the return air fan with those obtained in the occupied spaces, it is possible to evaluate if short-circuiting of the supply air is occurring.

The CO_2-based control for HVAC systems cannot be applied everywhere. One requirement of this approach is that it is only appropriate where people are the predominant source of air contaminants. This is because if the predominant source of air contaminants were from the furnishings in the building, or some activity or process in the building, then relying on just a CO_2-based ventilation control system might allow these contaminants to increase to irritating or unhealthy concentrations in the indoor air. While these conditions do not preclude the use of a CO_2-based ventilation control approach, they do indicate a need for a second evaluation criterion for the operation of the ventilation system, such as the measurement of respirable particulate matter or volatile organic compounds (VOCs). There should also be an absolute minimum of perhaps 0.1 cfm/ft^2 of OA.

While there are standards for not-to-exceed levels of CO_2, there has yet to be a consensus as to levels, particle sizes, and types of particulate matter. With respect to CO_2, achieving a minimum of 20 cfm of outdoor air per person in office spaces, as stipulated by ASHRAE Standard 62-1989, corresponds to not exceeding 875 ppm, assuming outdoor levels of 350 ppm. Please note, however, that if the outdoor levels were 400 ppm, then indoor levels would need to not exceed 925 ppm in order to achieve the same minimum of 20 cfm of outdoor air per person.

With respect to particulate matter, there is the *Federal Standard 209D, for Clean Rooms,* the least stringent of which is the 100,000 Class. This Class corresponds to achieving less than 100,000 particles per cubic foot for particles equal to or greater than μm in diameter. I personally consider this level of cleanliness an appropriate goal for offices spaces. As a point of information, the other more stringent Clean Room Classifications are 10,000; 1000; 100; 10; and 1 particles per cubic foot of air.

In addition to particulates, there are also VOCs that could be present in the indoor environment. VOCs are generated by such sources as organic solvents found on printing processes, paints, glues, varnish, cleaners, and perfumes. Formaldehyde is another VOC which can be emitted from glues used in paneling and carpeting, as well as from treated fabrics and new furniture. Particulates can arise from smoking, paper handling, or cleaning in the building, or they can be brought in from the outdoors due to inadequate filtration or infiltration.

The other requirements for the CO_2-based control strategy include the need for the HVAC system to have the ability to provide the adequate quantities of outdoor air (i.e., 100% economizer capability) to prevent the buildup of CO_2. There would be little benefit to installing this control equipment on a system that can only deliver a constant minimum quantity of outdoor air.

It should be remembered that the greatest benefit for the use of this approach would be in spaces with variable occupancy, such as classrooms, theaters, conference halls, meeting rooms, etc.; that is, locations where the HVAC systems would otherwise be continually providing fixed quantities of outdoor air based on the maximum potential occupancy. For instance, in a cinema, when the doors open at 1:00 p.m. and 20 people walk in and the operator simply turns the system on, the equipment ventilates that space for the maximum occupancy of 1000 people. This response fails to take advantage of the reserve of fresh air in this space at initial occupancy. By employing demand controlled ventilation, however, the cost of heating or cooling outdoor air when only a small amount is required represents a considerable potential cost savings for this and similar applications.

For those readers interested in pursuing this approach, there are already several CO_2 sensor/controllers of this type currently on the market. The manufacturers include (1) Gaztech International Corporation [6489-A Calle Real, Goleta, CA 93117, (805) 964-1699]; (2) Sauter Controls Corporation [5333 Mission Center Road, Suite 336, San Diego, CA 92108, (619) 291-1132]; and (3) Acme Engineering Products, Inc. [Trimex Industrial Building, Route 11, Mooers, NY 12958, (518) 236-5659] which offers several models.

Several articles have been written on how DCV works. In one article, Sodergren[3] compares three ventilation control strategies:

- Constant outdoor air rate
- CO_2-based ventilation control system
- Timing function

With the constant outdoor air rate, peak CO_2 concentrations exceeded 800 ppm in the afternoon. In comparison, the CO_2-based control prevented the CO_2 concentrations from exceeding the set-point of 700 ppm and yet reported an energy savings of 40% as compared with the constant outdoor air rate.

The reason for this improvement is that a constant ventilation rate only indirectly controls IAQ. The CO_2-based control approach permits a more direct control of air quality.

An understanding of the relationship between the HVAC controls and position of the OA dampers is very important for the performance of an IAQ evaluation. In order to be thorough, the amount of outdoor air delivered should be evaluated at the system's minimum outdoor air condition, as well as how the system is found to be operating. Therefore, the IAQ investigator or the staff provided to support the effort needs to be able to manipulate the system to make it go to its minimum outdoor air condition. In many systems, for instance, raising the setpoint of the mixed air sensor/controller will "trick" the system into going to its minimum outdoor air configuration.

VENTILATION DURING CONSTRUCTION

In addition to the ventilation systems that are intended to deliver adequate quantities of outdoor air to building occupants during the normal life of a building, there are special circumstances or conditions that require special provisions for ventilation. One of these special situations is when renovations or construction is going on in a building that is already partially occupied. The goal in this situation is prevent the air contaminants generated as part of the renovation/construction activities from being recirculated to the occupied areas of the building. These sources can include such things as dusts or solvents from carpet adhesives or paint fumes. In order to accomplish this, the zone of activity needs to be isolated from the rest of the building, and its air contaminants need to be exhausted directly to the outdoors. To do this, a temporary exhaust ventilation system should be installed. This installation may require the removal of a window panel in order to gain access to the outdoors. Also, the return ductwork serving this area should be sealed off, not only to prevent the redistribution of these air contaminants, but to prevent this portion of the HVAC system from absorbing these dust and VOCs and then releasing them back at some point in the future.

Another aspect of renovating or reconfiguring office spaces is that (all too often) it may be assumed that the existing ventilation will be adequate for the new loads, when in fact a particular building's HVAC equipment may *not* be adequate for the loads of the reconfigured space. In addition to the adequacy of the system's equipment, the locations of the thermostats, sensors, and other control devices should be reviewed. Several separate areas or enclosed offices on a single thermostat can result in poor thermal control and discomfort complaints, especially when their loads, either due to variations in occupancy or equipment densities, are not uniform.

Example 2.8. In one building, after renovation in a basement area which removed a wall, the thermostat for that zone was no longer in a sheltered corner but was now in a location being influenced by air coming down a stairwell from the first-floor lobby above. This cooler air at the thermostat led to overheating of this zone, with resulting occupant complaints and wasted energy. This wasted energy was due to the electric terminal reheat coils operating unnecessarily.

Another ventilation-related consideration for achieving good IAQ is the operation of the system during the initial occupancy interval, as pointed out by Levin.[4] Even the best ventilation system design and proper operation may be inadequate to remove contaminants shortly after installation of new building materials and office furnishings. Paints, flooring adhesives, duct sealants, insulations, foams, fabrics, and other materials generally emit at a much higher rate during the early period of exposure to the environment. To prevent IAQ problems in this situation, the following three steps are therefore recommended.

1. Provide extra ventilation during this initial period, especially if occupancy occurs immediately after construction and installation of furnishings. This extra ventilation can be achieved by increasing the outside air fraction to its maximum (100%, if possible) within the constraints of the heating/cooling capacity of the system.
2. Increase the number of hours that the ventilation system is operated. According to Levin, 24 hour per day operation, seven days a week makes sense for the "first 3 to 6 weeks of occupancy."
3. Operate the system at the lowest possible temperatures consistent with occupant comfort during this period. This helps reduce chemical emission rates, perceived air quality problems, and thermal discomfort.

A related issue, that of performing a bake-out involving elevated temperatures to drive VOCs out of building materials, has also been tested. While the potential of this approach has yet to be fully investigated, the initial indications are that it can create problems by driving the VOCs out of their original source materials into other materials which then become sources themselves. There is also the significant additional problem that these elevated temperatures will void the warranties of equipment, such as computers, that have already been installed in the building. Therefore, I would recommend against the performance of a bake-out in favor of selecting low-emitting materials and providing increased ventilation during and immediately after construction or renovation activities.

REFERENCES

1. Roberts, J. W., P.E. "Outdoor Air VAV Systems", *ASHRAE Journal*, pp. 26, 28–30. September 1991.
2. Gardner, T. F., P.E. 1988. "Part load ventilation deficiencies in VAV systems", *Heating/Piping/Air Conditioning*, pp. 89–92, 97, 100, February.
3. Sodergren, D. 1982. "A CO_2-Controlled Ventilation System", *Environment International*, Vol. 8, pp. 483–486.
4. Levin, H., Ed. 1990. *Indoor Air Quality Update*, Vol. 3, No. 5, p. 4. Cutter Information Corp. Arlington, MA.

CHAPTER 3

Individual Components of HVAC Systems

OVERVIEW

Before the performance of heating, ventilation, and air conditioning (HVAC) systems can be evaluated, it is necessary to first understand the specifics of the design of this system and the intended functioning of this equipment. This chapter presents a discussion of the individual component parts of HVAC systems, including an explanation of the relationship between each of these components and the achievement of good indoor air quality (IAQ).

Each component of the HVAC system has a specific importance to the resulting IAQ that the total system is capable of providing. This chapter discusses the role that the following components play in achieving good IAQ. As part of an inspection and testing effort, the condition of each of the components of the HVAC system needs to be considered. To facilitate this task, a checklist, such as the one presented in Figure 3.1, can be used. One of the challenges of performing IAQ evaluations is that to be thorough, all aspects of the system need to be looked at. The performance of the overall system is dependent on the proper functioning of each of the individual components of the system.

The following discussion of the relationship between each of these HVAC components and the achievement of good IAQ can be used in conjunction with an evaluation of the performance of a building and its systems. People already using another list of HVAC components, the "HVAC Checklist — Long Form" in the EPA/NIOSH document, *Building Air Quality*,[1] should find this discussion useful in understanding the importance and significance of the

Building: _____ Date checked: _____

Inspector: _____ Weather conditions: _____

Location of air handling units: _____

Type of HVAC System: CAV   VAV
Hybrid   Heat pump

Inspection Results/Comments

  1. Outdoor air intake
  2. Location of mechanical room
  3. Outdoor air dampers
  4. Mixing plenums
  5. Filter banks
  6. Face and bypass dampers
  7. Cooling coils
  8. Condensate drain pans
  9. Heating coils
10. Supply air fan
11. Humidification
12. Distribution system
13. Terminal equipment
14. System controls
15. Room partitions
16. Stairwells and elevator shafts
17. Return air plenum
18. Return air fan
19. Building exhausts
20. Boilers
21. Cooling towers
22. Chillers

Comments:

Figure 3.1   Inspection Report of Individual HVAC Components.

questions raised in that 14-page checklist and determining whether that component is OK or if it needs attention.

Outdoor Air Intake

When designing or evaluating a building, one of the first IAQ considerations that will affect the ability of the HVAC system to be able to provide good IAQ over the life of the building is the location of the outdoor air intakes. All too often, architects have adopted an "out of sight, out of mind" philosophy towards the air intakes in their buildings. By failing to recognize the importance of this component of the HVAC system, however, they may be doing the future occupants of their buildings a disservice. As a colleague of mine is known to say, "The outdoor air intakes are the nose of the building." This is an apt metaphor since the mechanical ventilation system must be able to ventilate the building, just as the human respiratory system must be able to ventilate the lungs. Human evolution has provided the nose with a prominent location on our anatomy; buildings should be treated similarly.

As part of an IAQ evaluation, therefore, the investigator should, where access permits it, stand in front of the outdoor air (OA) dampers and look around for sources of air contaminants. Potential sources will vary as a function of the outdoor air intake location. Examples of what might be found for rooftop air intakes include sewer vent pipes, cooling towers, kitchen exhausts, laboratory fume hood exhausts, or standing water. If it hasn't rained in a while, the roof surface should be inspected for indications that this condition has occurred in the past. Pouring water on the roof can assist in this determination. In one building investigation performed in the late winter, the remains of the previous season's growth of grass from between the rocks on the roof provided evidence of ponding. Ground-level air intakes are prone to be in the vicinity of vehicle exhausts. Potentially worse still are below-grade air intake wells serving basement mechanical rooms where bird droppings and decaying leaves can accumulate.

Air intakes located one third of the way up the side of the building tend to work best. According to systematic flow patterns around a large number of building shapes, Wilson[2] found that there is little mixing from the upper two thirds of the building with that from the lower one third of the building.

The investigator should also remember that the OA dampers represent only the location of the *intentional* outdoor air intakes. Frequently, there are also locations of *unintentional* air intakes; they can either be near the intentional air intakes or they can be elsewhere in the building. For instance, air is typically drawn into the air handling equipment directly from the mechanical room. The magnitude of this pathway will depend on both the leakiness of the equipment and the pressure differential between this equipment and the mechanical room, as well as between the mechanical room and the outdoors.

Unintentional air intakes located away from the mechanical room typically occur at the lower levels of the building and are due to the stack effect. The stack effect refers to the phenomenon where heated air, being more buoyant than the colder air outdoors, escapes at penetrations in the upper levels of a building and this leads to replacement air being drawn into the building at lower levels. The magnitude of this pathway will again depend on the availability of penetrations (i.e., the leakiness of the building) and the pressure differential created by the indoor/outdoor temperature difference and the height of the building. Examples of investigations into unintentional air intakes are presented in Chapter 8.

Another item to be considered when inspecting the location of the outdoor air intake, especially for sidewall installations such as with unit ventilators, is whether the OA grille has become obstructed, either intentionally due to past problems of coil freeze-ups, or unintentionally due to debris from nearby sources: traffic, trees, etc. Bird screens, with a minimum mesh size of $^1/_2$ in. are recommended because smaller sizes become obstructed too easily.

Mechanical Room

There are several aspects of the mechanical room that can affect the achievement of IAQ. These aspects include its location, its other uses, and its pressure relationship with respect to other areas. Recognizing that air handling units (AHUs) typically draw some air directly from the surrounding mechanical room, this location should be checked for its proximity to sources of air contaminants. I have performed more than one IAQ investigation where the AHUs serving office spaces were located in a mezzanine above a source of air contaminants. In one case, it was a Receiving Department and the source was vehicle emissions. In another, it was an airplane hanger and the source was emissions from the parked planes.

Other activities or equipment may be located in the mechanical room which can be a source of air contaminants. I have seen sumps with stagnant water in them that were not far from AHUs. Hazardous materials are also sometimes stored in mechanical rooms. Maintenance and repair activities with the potential to generate odors are also a concern. The condition of the AHUs should also be checked at this time, especially looking for evidence of leakage sites from the mechanical room in the AHU.

Outdoor Air Dampers

The OA dampers are the first critical component of the HVAC system in terms of being able to provide adequate ventilation for the building occupants. The basic requirement for providing ventilation when the building is occupied means that the OA dampers should be opened at this time. The only exception to this rule is during a "warm-up cycle" where the OA dampers remain closed for an interval of time at the start of the day when ventilation is not yet required because of diluting effect of the available volume of initially clean air. For this approach to be appropriate, the ventilation system will have had to have been operated sufficiently long the evening before to have purged the occupied spaces of air contaminants of human occupancy from the day before. Renovation and construction materials will continue to outgas for at least 3 to 6 months.

Unfortunately for the occupants, buildings are still being operated with their OA dampers closed when people are present in the building, beyond what is permitted in ASHRAE Standard 62-1989. There are several reasons why OA dampers remain closed despite the presence of people in buildings. Examples of these reasons include:

1. Time clock improperly set; in one case, it was 2 days out of phase, with Tuesday getting a Sunday schedule.

2. Linkage to OA damper not functional due to its absence or to damage.
3. Misdirected energy conservation effort on the part of a building operator attempting to minimize expenses.
4. Mistaken belief that closed dampers will leak and provide outdoor air equal to 10 to 15% of the total supply air.

With respect to assumptions of minimum amounts of damper leakage, this may have occasionally been true at some point in time; but for the dampers in buildings today, it is no longer a valid assumption. Many systems have two banks of dampers: one sized for providing the minimum outdoor air quantity (typically 15% of the total supply air volume) and the other sized for up to 100% of the fan capacity to utilize "free cooling" as part of an economizer cycle capability. In addition to evaluating if the minimum OA dampers are open, the linkages between the dampers and the mechanical actuators should be checked for the tightness of connections and ease of movement over the expected range.

The system controls should be manipulated to make sure that the dampers are operating as intended. In one recent investigation of a building with numerous individual heating and ventilating units, the system control logic was such that when the mixed air temperature was above the setpoint (55°F), the OA dampers went to their minimum position; and when this temperature was below this setpoint, the OA dampers went to full open. Adjusting this setpoint of this sensor/controller and observing the behavior of the dampers identified one unit which stayed in its minimum outdoor air position no matter what the setpoint was. It was also determined that this unit was only providing about 2% outdoor air in this condition, which meant all of the time.

There are two basic types of controllers for OA dampers, and other dampers as well. There are modulating dampers that can vary the position almost infinitely between "full open" and "full closed," and there are controllers which will position the dampers only between the "full open" and "fully closed" position. In this latter case, modulation of the quantity of outdoor air is achieved over time, as the dampers are repositioned as often as once a minute, as the system "hunts" for the needed mixture of OA and return air (RA).

Mixing Plenum

This location is where the return and the outdoor air streams are combined together. The ratio of this mixing is a function of the positions of both the RA and the OA dampers and the quantity of relief or exhaust air. In some large systems, this portion of the HVAC is actually being used as a storage area for all sorts of things, some of which can be sources of air contaminants. This location should also be checked for cleanliness.

Since rain and snow may be sucked in along with the outdoor air, there is often a floor drain so that water does not accumulate. If this floor drain is connected to a sewer line, it needs to have a trap in it, with water present, so that sewer odors will not be drawn into the air stream.

Dampers, for the outdoor air, return air, and exhaust air should be checked for air-tightness. For instance, if the RA dampers don't seal effectively, there will always be a certain amount of recirculation of building air occurring. Dampers won't function unless the related actuators, linkages, and controls are in proper working order; so these should be checked as well.

If the system has a mixed air temperature control setting, its setting should be verified. If it is seeking too warm a temperature, it may not be letting in enough outdoor air. The freeze stat setting should also be documented because if it is too high, it can needlessly cause the system to shut down for freeze protection of the coils.

Filter Banks

The condition and quality of the filtering media and its installation are a very important determinant of HVAC system ability to achieve good IAQ. This is of course because of the requirement that the HVAC system deliver clean air. An inspection of the installation should first determine whether they are grossly clogged, blown out, and how completely they fill the cross-section of the air handling equipment. This absence of gaps is critical for the rated effectiveness of the filters to be achieved. Compared with the high-pressure drop across the filter media, the lack of resistance to airflow at any gap will permit a volume of air to leak through that opening at a rate much greater than the ratio of that area to the total area. That is, even a small gap between the filters will permit a large volume of air to bypass the filters.

In addition to the condition of the filters themselves and their installation, others issues of interest include: (1) what is the efficiency of these filters?, (2) how often are the filters changed?, (3) how is the decision to change the filters made?, and (4) is there a log of these filter changes being kept?

With respect to the efficiency of the filter media, what is the minimum acceptable efficiency? The answer will depend on the goal. According to Morey,[3] "a 50% atmospheric dust spot efficiency filter will remove most microbial particulate that may be entering in the return or outdoor air streams." At a minimum, there should be 1 to 2 inch thick extended surface (i.e., pleated) filter that is rated at least 20 to 30% by the dust spot method.

Filters are rated according to the requirements of *ASHRAE Standard 52-76*[4] to yield a *dust spot %* and an arrestance %. Of these two criteria, the *dust spot* is a more meaningful indicator of efficiency. The *arrestance* merely refers to a percent removal on a weight basis. Since the larger, easier-to-collect particles will represent most of the weight of particulate matter, filters with low dust spot ratings can have high arrestance values. For instance, according to their literature, a Farr 30/30 filter, which is considered a medium-efficiency filter with a dust spot efficiency of 30%, has an arrestance of 94 to 96%. Another aspect of media filters is that their collection efficiency increases over time as the filter cake builds up. According to Farr,[5] the efficiency of their RIGA-FLO® 10 and XL filters, a

0.3 μ diameter particulate, starts at 5% and increases to 55%, with a weighted average of 34%. These filters are rated at a 40 to 45% dust spot and a 96% arrestance. Similarly, their RIGA-FLO® 100 filters have a 0.3 μ efficiency that starts at 48% and increases to a final value of 86%, with a weighted average efficiency of 68%. These filters are rated at an 80 to 85% dust spot and 98% arrestance.

If high-efficiency, extended media filters are changed merely according to a certain time interval (e.g., once a year or four times a year) there is the risk that the filters will be changed too often, which is wasteful of finite resources, or they will not be changed often enough. This determination cannot be made accurately by merely a visual determination because, unless they have been changed recently, properly functioning filters will look dirty on their upstream side. As a filter loads up, the pressure drop across it increases, the filtration efficiency also increases, but the volume of volume passed through decreases. The magnitude of this decrease in volume is a function of the characteristics (i.e., fan curve) of the fans in the system and of the filter design. The limiting factors in this situation include both meeting the required minimum air volumes and the structural integrity of the filter frames. The problem is that a point is reached where the frame of the filter cannot maintain its shape under the increased pressure drop, so it deforms, bending in on its sides, thus creating gaps for air to bypass the filters without benefit of cleaning. The appropriate procedure to follow then, is to use the manufacturer's recommended final resistance, monitor the pressure drop across the filter bank, and then change the filters when this ΔP (Δ = Greek letter delta, meaning differential) has increased to where the manufacturer recommends replacement. In order for this procedure to be followed, there needs to be functioning pressure gauges in place to measure the pressure drop across the filters.

The condition of this detail says a lot about the level of maintenance provided in a building. I have seen systems devoid of any pressure gauges, or just the dusty relics of gauges but most of the tubing gone. I have also seen systems where there are spotless magnehelic gauges accompanied by logs of the pressure drops when filters were replaced, along with the date and the initials of the individual responsible. Therefore, the inspection should check for the presence and condition of pressure-sensing equipment, as well as for the presence of gaps around the filters. Also, check for dampness of all filters, especially any that are downstream from humidification devices, because the combination of this dampness and the collected dirt on the filters will cause the proliferation of microorganisms on the filter.

In addition to the increase in the pressure drop across the filters, another criterion for their replacement is when the filters themselves become a source of odors. As discussed in greater detail in Chapter 10, heavily contaminated filters exposed to relative humidities of 75% were shown to generate odors associated with the volatile metabolic products produced by microbes.[6]

Filters are now starting to include the presence of activated carbon, or other sorbant such as aluminum oxide, in an attempt to control odors as well as particulate matter. There are several issues involved with this effort. Sometimes carbon impregnated filters are suggested to eliminate problems arising from an outdoor air intake that is being impacted by odors from process exhaust or loading docks. It needs to be realized that this approach is just treating a symptom of the problem and not addressing its cause. There also remains the question as to when the carbon has become saturated and is therefore no longer providing any benefit. Unlike particulate collection, where the increase in pressure drop across the filter provides a criterion for when to replace the filter, there is not yet an easy way to make this determination for the remaining capacity of sorbants.

Face and Bypass Dampers

Dampers can be used to bypass a coil, such as an OA preheating coil, to provide thermal control or to eliminate unnecessary pressure drops in the system. In addition to checking that the operation of these dampers and their motors are correct, it is important to check for additional locations where outdoor air is drawn into the unit. In one installation, the bypass dampers were located on the bottom of the unit, very near to standing water.

Cooling Coils

In order to provide thermal conditioning in a building, the HVAC equipment either adds heat to or removes heat from the supply air stream. This can be accomplished with either one coil, which is changed between cooling and heating depending on the building demand, or multiple coils which are dedicated to either heating or cooling.

The cooling coils of the HVAC systems can use chilled water, brine, glycol, or various types of refrigerant as the cooling medium. Coils cooled by refrigerants can be either of the direct expansion (DX) or flooded types. In terms of IAQ, one aspect of the system design to consider with respect to cooling coils is the amount of cooling that can be achieved in the building in water-based systems. One measure of the capacity of the coil is the temperature of the coolant entering and leaving the coil. Design cooling capacities were calculated based on being able to compensate for the anticipated heat sources (internal gains) assumed to be present in the building. These calculations are based on assumed numbers of people, equipment, lights, and solar gains, all of which generate heat in the building. While solar gains have not changed over time (with the possible exception of the removal of a structure that had been shading the building under evaluation) and the amount of heat generated per person has not changed, the density of these people can be increased over time, and perhaps most significantly, the use of heat-generating equipment has typically increased as computers and copiers have proliferated in the modern office environment. It is therefore possible that the original design capacity of the cooling coils in older buildings is not sufficient to maintain thermally comfortable conditions throughout the building. Overheating can be considered to be a degradation of the IAQ. Therefore, the temperature of the air leaving the coils and the temperature of the delivered supply air as well should be considered with respect to its ability to provide adequate cooling. In one building investigation, the design called for an exit temperature from the cooling coils of 53°F. The temperature observed, however, was 64°F. This situation contributed to complaints of overheating at various locations in the building.

In terms of building IAQ, another aspect of the cooling coils to consider is the ability of the system to remove the water condensed out of the air stream. As air is cooled, its ability to maintain moisture in its vapor phase is decreased. There-fore, the relative humidity of the air stream will rise. If the incoming air is cooled to below its dew point, the relative humidity will reach 100% at the coils and some of the moisture will condense into liquid droplets. There was a time (when energy was much cheaper than it is today) when HVAC systems were designed to overcool the air down to a desired absolute humidity level and then reheat this air back up to the desired delivery temperature. This practice has all but been eliminated as being wasteful of energy, except in areas requiring precise control of humidity. As a consequence, however, the air leaving the cooling coils has a high relative humidity, frequently approaching 100%. If the ductwork down-stream of these coils is dirty, then the potential exists for the growth and ampli-fication of microbiological organisms in this ductwork. While clean fiberglass duct lines will not absorb sufficient moisture to support microbial growth, an accumulation of dirt on the surface of the fiberglass liner will increase the amount of moisture absorbed to the point where growth will be fostered.

Another aspect about cooling coils that can affect the IAQ achieved in the building is that these coils have a maximum permissible face velocity (volumetric air flow rate divided by the area of the face of the coil) beyond which the condensing water droplets will be carried over in the air stream. This condition will increase the potential for the creation of an environment in the ductwork hospitable for the growth of microorganisms. One reason why the face velocity of the cooling coils gets exceeded is that the supply air fans have been sped up beyond their original design condition in order to deliver more air to the building. This situation frequently occurs in response to a need to provide more cooling due to the presence of increased sources of heat generation (such as computers on almost every desk) when the original design calculations were only for people and perhaps limited use of electric typewriters. Fortunately, however, energy efficient lighting can help lower cooling loads because fewer watts of electricity are required per lumen of illumination. Another reason why cooling coils have been found to be exceeding their rated face velocities is that a smaller, and therefore cheaper, cooling coil was substituted for the equipment actually specified, either as a last-minute attempt at cost saving or by mistake.

Another way a problem with cooling coils can adversely affect the achieve-ment of adequate IAQ is if the coils are dirty. A layer of dirt on the surface of the coils can reduce their ability to cool the air passing over them, effectively reducing their cooling capacity. If the coils are dirty enough, this will increase the pressure drop across them and further reduce the overall system cooling capacity by reducing the volume of supply air being delivered.

Condensate Drain Pan

The water condensed out of warm, humid air by the cooling coils needs to be drained away from the tray under the coils. There are several requirements that are necessary for this drain pan to empty completely of water under all operating conditions. The importance of preventing standing water in the HVAC system is crucial for preventing the growth of microbiological contamination within the HVAC system. It is therefore imperative for the maintenance of good IAQ that the water condensed by the cooling coils be allowed to siphon out of the pan of a draw-through system. If not, the resulting pool of stagnant water will cause the growth of microorganisms in the HVAC equipment, downstream of the filters, where the spores released can directly affect the people in the building. In order to prevent water from accumulating, several aspects of the equipment need to be properly designed, installed, and maintained. In terms of design, there needs to be a properly designed water trap in the drain line, between the AHU and the mechanical room, which will isolate the pressure differential from inside the equipment to outside the equipment and allow the condensed water to be siphoned out of the pan. In terms of installation, after checking for the water trap, the slope of the drain pan needs to be checked, as well as the connection of the drain pipe to the drain pan. If the drain pipe is not connected to the lowest point of the pan, then water will collect. In terms of operation and maintenance of the system, the drain pan and trap need to be free of any clogs or debris. The details of this installation are summarized in Figure 3.2. If an inspection is performed, the following items need to be checked:

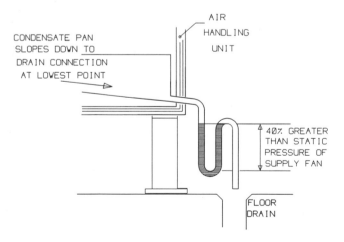

Figure 3.2 Detail of water trap at condensate drain line.

1. The ease of access to permit inspection and cleaning.
2. The cleanliness of the drain pan. This should be obvious as part of the requirement for any evidence of past or recent contamination. Also, debris can block off the drain pipe and lead to standing water.
3. The connection of the drain pan to its drain pipe. This pipe needs to be connected to the pan at the lowest point of the pan so that all of the water will be able to drain out. If the bottom of the drain pipe is above the bottom of the drain pan, then water can accumulate and remain stagnant. The drain pan also needs to have a definite slope *towards* the drain pipe.
4. The presence of a water trap in the drain pipe. Inasmuch as the cooling coil is typically located between the filter bank and the supply air fan, there will be a large negative pressure between this portion of the AHU and the room air of the mechanical room. If there were not a water trap in this drain to isolate this AHU from the mechanical room, then the air being sucked into the AHU from the mechanical room could prevent the water from draining away. To be properly designed, the effective height of the water trap should be 40% greater than expected peak static pressure (in inches of water column) of the supply air fan.

Heating Coils

The issues involved with heating coils are much the same as those for cooling coils: the need for access, the need to be kept clean, and the need for them to have sufficient capacity in order to provide thermal comfort in the occupied spaces. Here, they need to be able to provide a sufficient amount of adequately heated outdoor air during the winter design conditions.

Supply Air Fans

The supply air fan or fans provide the driving force to move the air through the distribution system. There can be just a supply air fan or the combination of a supply air fan and return air fan in the HVAC system. There will be, of course, separate exhaust fans for the toilet exhaust and any other identifiable sources of air contaminants, such as from a kitchen. An inspection of the supply air fan should make sure the fan is operating and include a check of the condition of the fan belts, the fan housing, and any louvers which could be restricting the flow of air into the fan. If there is a return air fan as well, this fact should be noted and its condition checked as well. A more detailed evaluation may also need to consider the relationship between the operating parameters of the supply air fan and those of the return air fan in order to understand the pressure relationship in the building. For instance, if the return air fan is attempting to move more air than the supply air fan, or if the building exhausts exceeds the capacity of the supply air fan, then the building will be operated at a negative pressure with respect to the outdoors instead of at a positive pressure. (See Chapter 8 for discussion of building pressurization.)

Humidification

In addition to the removal of moisture from the air stream, the introduction of moisture to the air stream represents a potential source of air contaminants in HVAC systems. The absence, or the lack of the operation, of humidifiers can adversely affect the comfort of individuals in buildings located where the outdoor conditions are cold and dry. *ASHRAE Standard 55-1981*[7] defines acceptable ranges of operative temperature and humidity (see Figure 3.3). People feel most comfortable if the indoor air relative humidity is between 30 and 70%. Lower humidities will dry out the mucous membranes of the building occupants; contact lens wearers will complain of discomfort trying to wear their lenses all day; and

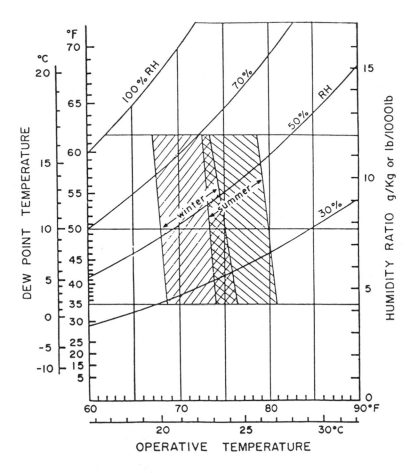

Figure 3.3 Acceptable ranges of temperature and humidity for a person clothed in typical summer and winter clothing engaged in light activity (mainly sedentary, <1.2 metabolic equivalents). *Source: ASHRAE Standard 55-1981.*

this dryness will permit air contaminants such as particulates or volatile organic compounds (VOCs) to be more irritating.

The presence of humidification, however, is no guarantee of good IAQ either. Problems can arise from contaminants in the water supply, or the accumulation of moisture in an inappropriate location can be a contributing factor that permits microbiological growth within the HVAC system. The water source for humidification should only be from potable quality sources. Some of the chemicals used in boiler water treatment compounds are hazardous to health. Examples of compounds that can be found in boiler water that should not be used in direct-steam humidification systems include cyclohexylamine, diethylaminoethanol, and morpholine.

During the inspection of the humidification equipment, it should be noted that if standing water is present, it can serve as an excellent medium for the growth of microorganisms. This device can therefore act both as a source of contamination and also as its amplifier. Therefore, according to the ACGIH Committee on Bioaerosols,[8] the following actions are recommended to prevent the growth of microorganisms.

1. Remove stagnant water and slimes from building mechanical systems. When the system is off, disinfect with a chlorine solution.
2. Eliminate water spray systems in air handling units.
3. Substitute steam for cold-water humidifiers.
4. Eliminate the use of cold-mist vaporizers.
5. Prevent leaks or floods.
6. Keep indoor relative humidities below 70%. For locations where cold surfaces are in contact with warm air, keep indoor relative humidity below 50%.
7. Discard porous building materials and office furnishings that are visually contaminated with fungi.
8. Replace HVAC system filters at scheduled intervals.

If the inspection indicates that any of these recommendations are not being followed, then this observation should be included in the report. Both the control of humidity in indoor environments and proper maintenance of humidifying equipment are important preventive strategies for achieving good IAQ. In colder climates, any attempt to provide humidification should first include a determination as to whether or not the details of the building envelope were specifically designed and built so as prevent condensation on surfaces. The amount of humidification will then also be a function of the expected minimum outdoor temperatures.

The Distribution System

The role of the distribution system, which consists of the ductwork, mixing boxes, and terminal units, is to convey the conditioned air from the air handling equipment throughout the building to where the people are, without adding any air contaminants to this air stream. Therefore, the concerns about the condition of these components of the distribution system, in terms of maintaining good IAQ, relate to both the physical integrity of the ductwork and its connections, plus the condition of the interior surfaces of these components.

In terms of the integrity of the system, a visual inspection of representative portions of the ductwork and distribution boxes should be made. Things to look for include the presence or absence of duct sealing caulks and the presence or absence of dirt at corners of the ductwork. This last item is mentioned because one indication that air is leaking directly from the supply ductwork to the return plenum is that dirt in the supply air, which has bypassed the filters, gets deposited in the vicinity of the leak.

Example 3.1. In one building investigation, the complaints included not only concerns about the quantity of fresh air, but also thermal comfort issues in terms of frequent overheating. The overheating suggested that perhaps an adequate quantity of supply air, as well as outdoor air, was not delivered to the zone served by this particular rooftop AHU. The first determination from the visual inspection of this AHU was that the OA dampers were fully closed during occupied intervals in the building. Peering more deeply into the supply air ductwork, it was then observed that a seam of the fiberglass ductboard directly under this AHU had burst and was therefore permitting most of the supply air to short-circuit directly back to the suction side of the fan without reaching the occupied space.

Beyond visual inspection, the techniques for quantifying the amount of leakage from the distribution system is presented and discussed in Chapter 6.

In addition to the integrity of the connections of the ductwork and other components of the air distribution system, the condition of the interior surface of these components is also important in terms of the IAQ being provided by that system. A thorough IAQ evaluation should include a visual inspection of portions of the interior of the ductwork, especially where interior sound-absorbing liners are expected to be present. One location to be checked is the ductwork closest to the supply air fan. The surface of the sound-absorbing duct liners at this location frequently become eroded over time. In a paper presented at the *Healthy Buildings — IAQ '91 Conference,* a study of the porous insulation of five office and commercial buildings was cited.[9] This paper reported that it is almost impossible to prevent the accumulation of dust and debris in insulation pores. The suggested remedial measures include efforts to limit the availability of nutrients and moisture, as well as rigorous preventative maintenance in order to prevent microbial contamination in buildings. One manufacturer of packaged AHUs is now offering units with double-wall construction which prevents exposure of the fiberglass insulation to the airstream.

The magnitude of the porosity problem with insulation is sometimes aggravated when erosion by the air stream damages the insulation liner. The additional ingredient to the dirt not trapped by the filters that ends up in the porous insulation is the *high relative humidity of the air leaving the cooling coils*. This combination of dampness, dirt, and darkness are the basic ingredients for the growth and amplification of microorganisms. A more detailed discussion of these air contaminants is presented in Chapter 10. In addition to concerns about the potential for microbiological growth, the existence of eroded duct liners means that fiberglass particles, a respiratory and skin irritant, are being scoured from this surface and delivered to the breathing zone of the occupants in the building. Deteriorating duct liners should therefore be removed from the inside of ductwork. Insulation against condensation, which includes a vapor barrier, can be replaced on the outside of the ductwork. Insulation against condensation, which includes a vapor barrier, can be placed on the outside of the ductwork. Other installations, which relied on the interior fiberglass for noise control are now using metal sound attenuators.

In addition to the lined ductwork, terminal boxes are also frequently lined as well. The condition of these liners should also be inspected. In response to this problem, one manufacturer of terminal boxes now offers an optional "Steri-Loc"[10] liner with a "smooth, nonporous skin to guard against mold, spores, bacteria and air erosion." The other concern about erosion of duct liners, in addition to the microbiological habitat created, is the fact that this material is being introduced into the supply air downstream of the filter bank and can therefore be distributed to the occupied areas of the building. When this condition exists, the occupants will typically be very unhappy about this fact. Pieces of this eroded liner material can also clog the coils of terminal reheat devices which then reduces the volume of supply air that can be delivered to locations downstream of those devices.

While the interior of ductwork or its liner is never sterile, it should not contain a thick layer of deposited material. In terms of a visual inspection, metal-lined ductwork should still look metallic (i.e., should not be covered with dirt). Fiberglass-lined ductwork should be smooth inside, not worn, and should be its original color, not black, grey, or moldy blue-green. All ductwork should be completely dry. It is sometimes useful to check the relative humidity within the ductwork. Relative humidities in excess of 60% are at increased risk of promoting the growth of microorganisms.

Terminal Equipment

In addition to concerns about the clogging of coils in terminal reheat boxes, terminal equipment with the potential for adversely affecting IAQ include induction fan-coil units and fan-powered terminal boxes.

Induction units use the movement of supply air to induce the movement of air in the space to achieve a larger combined total airflow. Induction units are usually under a window or at a perimeter wall, with centrally conditioned primary air supplied to the induction nozzles of the unit at high pressure. This flow then induces secondary air from the room through the secondary coil. Primary and secondary air is mixed and discharged to the room. The secondary air is either heated or cooled at the coil, depending on the season, the room requirement, or both. This coil has a drain pan to collect any condensed moisture that accumulates during cooling. A lint screen is normally placed across the face of the secondary coil. The potential IAQ problems associated with this equipment relate to the combination of the situation where, with many of these units distributed in the building, the coils and the condensate pans located at floor level may not be cleaned as often as needed. I have also seen perimeter fan-coil units where the condensate tray, although not at floor level, was very difficult to access for either inspection or cleaning and therefore was not inspected or cleaned. The resulting dirt in the pan can trap moisture and lead to microbiological contamination. Another microbiological problem has been observed with these units when the valve controlling the heated or chilled water leaks and wets the adjacent carpeting.

Another type of terminal unit, a fan-powered mixing box, can also be a potential source of degraded air in those buildings where they are used. As discussed in Chapter 2, these fan-powered units can recirculate air from the local return plenum to perimeter locations to provide supplemental heating. The potential problem, in terms of IAQ, is that many of these units were not set up to achieve good filtration of the air from the return air plenum. This is because return air plenums are not kept clean. The operation of these fan-powered terminal boxes can then become a conduit for relocating particulate matter from the return plenum to the occupied spaces. The other aspect of these terminal units is that, like distributed perimeter unit ventilators, it is a challenge to provide adequate maintenance to these units. Not only are they distributed throughout the building, but access to them can be difficult, depending on the type of suspended ceiling and their location with respect to the current floor plan, which may have been modified from its original layout.

Another HVAC system component classified as terminal equipment is a variable air volume (VAV) box. As also discussed in Chapter 2, this device regulates the quantity of supply air delivered to a particular zone. If this zone is not calling for cooling, or heating, even when it is occupied, then it will restrict the quantity of air delivered down to some minimum value. This minimum delivery of air may therefore not deliver an adequate quantity of outdoor air.

System Controls

The relationship between the HVAC system controls and the achievement of good IAQ was presented in Chapter 2. In terms of evaluating the role the controls are playing in the nonachievement of good IAQ, this can be a very difficult situation. Unlike the presence of a dirty or nondraining condensate pan, the specifics of the control functions are far more subtle and more difficult to decipher. As mentioned in Chapter 2, there was one building in which a mixed air sensor/controller was out of calibration, leading to the OA dampers being closed.

In another building, a thermostat was located in the return plenum instead of the occupied spaces. This in itself was not a problem except that in this system, much of the supply air was leaking directly to the plenum without reaching the occupied spaces. This resulted in the system not being able to control for the conditions actually occurring in the space.

Another potentially difficult to interpret aspect of controls are pneumatics, where the air pressures required to perform different functions vary according to the design intent of the system and also the control manufacturer. For a typical system with a supply pressure in the range of 15 to 20 psi (pounds per square inch), the sequence of operations might be as follows: the outside air and exhaust air dampers start to open at 5 psi. The recirculating damper starts to close at 8 psi. The outside air and exhaust dampers are fully open at 10 psi, while the recirculating dampers are not fully closed until the control output pressure reaches 13 psi. The arrangement allows for a smooth transition when the dampers are modulating between return and fresh air.

Room Partitions

Room partitions can affect IAQ by interfering with the distribution and mixing characteristics of the supplied ventilation air. Many IAQ problems have resulted when an area which was originally intended as an open area is reconfigured to include partitions. The supply diffusers present will have been selected to achieve adequate distribution without obstructions and may no longer be able to deliver air to where the people are. Particularly troublesome in this respect are partitions that fail to leave a gap at their bottom. The presence of the bottom gap can improve the situation by permitting a convective airflow pattern to exist, with the heat given off by the people and equipment driving this airflow. For the partitioning of open areas to smaller spaces, one needs to make sure that the distribution of supply and return registers is still appropriate for the desired air circulation requirements.

Return Air Plenum

Another portion of the distribution system is the return plenum which provides the pathway for the loop to be completed for the recirculated air. The penetrations between the occupied spaces and the return plenum and the associated pressure differentials can affect the distribution patterns in the occupied spaces. For instance, all ceiling tiles must be in place for a plenum system to perform as intended. Missing ceiling tiles can redirect the return airflow, depriving more distant locations of their intended quantities of supply air. Dirt deposited on the ceiling near the diffusers usually consists of carbon particles, but occasionally could be mold. This determination can be made with the use of a good microscope. If the dirt build-up is large, the cause should be determined. In one case, the source was identified as due to excessive wear on the fan belts.

Another inspection item with respect to the ceiling tiles is the presence of water stains. In a well-maintained building, there should not be any ceiling tiles with moisture stains. If ceiling tiles do become damaged by water leaks, after the cause has been identified and corrected, the damaged tiles should be replaced with new ones since the top of the tiles often constitute the bottom of the return air plenum and they could be a significant source of microbial contamination.

Example 3.2 (Extreme). In one poorly maintained building, when I went to investigate water dripping from a ceiling tile, I narrowly escaped serious injury when the bucket that was resting on the ceiling tiles to collect the leaking water crashed to the floor through the soggy tiles. This is another example of maintenance treating symptoms but not addressing the cause.

Example 3.3. In another building, the walls between offices did not stop at the bottom of the suspended ceiling, but continued all the way up to the top of the plenum. The intent of this was to improve acoustical privacy between adjacent offices. The presence of the walls in the return plenum, however, acted as restrictions to the free flow of air and reduced the total quantity of supply that could be delivered to the space. That is, since the return plenum is an integral part of the air distribution system, an obstruction here affects the performance of the whole system.

The distribution of pressure differences between the return plenum and the occupied space will also affect the distribution of the supply air in that occupied space. In one building without return air ductwork and where the mechanical room was open to the return plenum, there was a pressure gradient across the suspended ceiling which decreased as a function of distance away from the mechanical room in the core of the building. This resulted in reduced air circulation at the perimeter areas.

Where return air ductwork or plenums pass through firewalls, they are required to have fire dampers. These locations should therefore be inspected to check if these dampers have closed, if there are concerns about the adequacy of the total supply air volume being circulated.

Since the return air plenum will be at a negative pressure with respect to the outdoors, as well as to the occupied spaces, penetrations between the plenum and the outdoors will function as unintentional outdoor air intakes. This can result in the introduction of odors or other air contaminants and can upset the pressure relationships in the building. If a large enough volume of air is entering the building via this pathway, then the intended outdoor air intake may actually end up functioning as a building exhaust.

Stairwells and Elevator Shafts

Stairwells and elevator shafts become important in terms of the achievement of good IAQ because of the pathways of air movement that they can provide through the building. The actual movement of air will be a function of the pressure relationships present. As discussed in Chapter 8, these pressure relationships may be due to the stack effect due to heated air in the building, or due to negative pressures created by leaks in the air handling equipment. As pathways of air movement, these components of the building can facilitate the transmission of odors from one location to another.

Example 3.4. In one three-story building with heat pumps above the suspended ceiling of each floor, doors were propped open on one floor that was being painted in order to increase the amount of ventilation being provided. Normally, this type of system would only recirculate air on each floor. The propped-open doors, however, facilitated the paint odors being distributed to the other floors.

Example 3.5. In another building, the elevator shaft created a conduit for the transfer of air contaminants from a basement activity to the offices on the top floor of the building.

Return Air Fan

The function of the return air fan is to draw air out of the occupied spaces and deliver it back to the AHU. Air leaving the return air fan can also be exhausted directly to the outdoors. As included in the discussion of supply air fans, and again in Chapter 8, there is a critical relationship between the return air fan and the supply air fan.

Building Exhausts

The quantity of outdoor air entering the building will be exactly equal to the quantity of air leaving the building. Therefore, in the absence of either fan-powered exhaust at the AHU or pressure relief fans in the building, the upper limit to the amount of outdoor air that can be brought into the building will be equal to the sum of the building exhausts and exfiltration. Therefore, in buildings with only limited amounts of exhaust capacity, the amount of outdoor air for ventilation may be similarly limited and therefore inadequate.

Example 3.6. In one building, where there were rooftop packaged AHUs, the building relief fans were an option; thus, they required field installation, as opposed to the rest of the AHU equipment which was prewired at the factory. Upon inspection, it was determined that the building relief fans had not been wired correctly, so they never became operational and therefore there was an inadequate quantity of air for ventilation being provided to the occupants of the building.

For buildings with large amounts of air being exhausted, an IAQ problem may result from an insufficient quantity of make-up. Here, the occupied spaces of the building will be negatively pressurized with respect to the outdoors. This will result in the introduction of unfiltered and unconditioned air being drawn in at perimeter leakage sites in the building envelope. There is also the potential that some of the exhaust air will be reentrained and reenter the building.

Boilers

Boilers which generate the heated water for thermal conditioning may burn fossil fuels or use electricity. The burning of fossil fuels generates combustion products which can become air contaminants if they have a pathway to the occupied spaces of the building. The potential seriousness of this potential problem needs to be appreciated.

Example 3.7. In late winter of 1991, several people were hospitalized because of CO poisoning resulting from the leakage of boiler flue gases to guest rooms at a hotel in Florida.

Example 3.8. As part of an IAQ investigation of an elementary school with heating and ventilating units that consisted of residential gas-fired furnaces, it was determined that downdrafting of combustion by-products in the furnace room was occurring. This was occurring because someone had placed a fiberglass batt over most of the open area of the louver in the door to this room in order to reduce the likelihood of freezing the water supply pipes which shared this location. Since air from the furnace room was also being drawn into the furnaces at the fan housing, this created a pathway for the transport of the products of combustion into the occupied zone.

Cooling Towers

There are several relationships between cooling towers and the IAQ in buildings that are important. One relationship is a function of the cooling capacity of the cooling tower in rejecting heat to the atmosphere, which in turn is a controlling factor in the cooling capacity for the building. The second area of importance to be evaluated when considering the potential contribution of cooling towers to degraded IAQ is that water and dirt that can accumulate in the tower can be a suitable habitat for the proliferation of microorganisms. Therefore, the distance between cooling towers and air intakes should be reviewed for the potential of mist from the tower to reach the OA intake. Cooling towers also require a plan for appropriate chemical treatment.

Chillers

Chillers have an indirect role in the maintenance of good IAQ. In addition to being able to provide sufficient quantities of chilled water to meet cooling loads necessary for providing thermal comfortable conditions, the improper operation of this equipment can adversely affect the quantity of outdoor air entering the building.

Example 3.9. In one building with two large chillers, there was the operational requirement of maintaining a minimum load on the chiller to prevent the exiting water temperature from dropping too low. In order to keep the necessary load on the chiller, the HVAC was operated so that cooling the recirculated air was accomplished entirely by the cooling coils. This mode of operation kept the OA dampers at their minimum position, even when the outdoor air temperature would otherwise be conducive for the economizer mode and the resulting "free cooling" and increase the quantity of ventilation air being provided. This situation is a good example of the need to fully understand the details of the operation of all components of the HVAC system when performing an IAQ investigation.

Example 3.10. The temperature of the chilled water available from the chiller can also have an impact on the achievement of good IAQ. In one building, the original design called for 45°F chilled water. The system as found, however, was only providing 55°F chilled water. This limited the amount of cooling and dehumidification that the HVAC could achieve and contributed to the proliferation of microorganisms in the building due to the excessive humidities.

REFERENCES

1. U.S. EPA, 1991. "Building Air Quality: A Guide for Building Owners and Facility Managers." EPA/400/1-91/033, DHHS (NIOSH) Publication No. 91-114. S/N 055-000-00390-4. Available from Superintendent of Documents, P.O. Box 371954, Pittsburgh, PA. 15250-7954.
2. Wilson, D. J., and R. E. Britter, 1982. "Estimates of Building Surface Concentrations from Nearby Point Sources," *Atmospheric Environment,* 16:2631–2646.
3. Morey, P. R., 1988. "Microorganisms in Buildings and HVAC Systems: A Summary of 21 Environmental Studies. Engineering Solutions to Indoor Air Problems." *Proceedings of the ASHRAE Conference,* IAQ 88, Atlanta, GA. pp. 10–24.
4. *ASHRAE Standard 52-76, Method for Testing Air-Cleaning Devices Used in General Ventilation for Removing Particulate Matter.* 1976. American Society of Heating, Refrigeration and Air Conditioning Engineers, Inc., Atlanta, GA.
5. *Filtration and Indoor Air Quality: A Two-Step Design Solution.* 1992. Farr Company, El Segundo, CA. 36 pages.
6. Pasanen, P. et al. 1990. "Emissions of Volatile Organic Compounds from Air Conditioning Filters of Office Buildings." *INDOOR AIR '90: The 5th International Conference on Indoor Air Quality and Climate.* Toronto, Canada. pp. 183–186.

7. American Society of Heating, Refrigeration, and Air-Conditioning Engineers. 1981. *ASHRAE Standard 55-1981: Thermal Environmental Conditions for Human Occupancy.* Atlanta, GA. ASHRAE.

8. American Conference of Governmental Industrial Hygienists (ACGIH) Committee on Bioaerosols. "Bioaerosols: Airborne Viable Microorganisms in Office Environments: Sampling Protocol and Analytical Procedures." *Applied Industrial Hygiene* (1) April 1986. pp. R-19 to R-22.

9. *ASHRAE Journal,* October 1991, p. 7.

10. Titus Advertisement for Steri-Loc™ liners appearing in the *ASHRAE Journal,* September 1991, p. 4.

CHAPTER 4

Evaluation Criteria for Indoor Air Quality

OVERVIEW

As described in the Chapter 1, indoor air quality (IAQ) problems arise when there is an inadequate quantity of ventilation air being provided for the amount of air contaminants present in a given space. Therefore, *standards* or *guidelines* attempting to regulate IAQ can do so by specifying minimum quantities of ventilation air, by specifying maximum concentrations of air contaminants that are allowable, or by specifying both. Therefore, this chapter is divided into discussions of both approaches. With respect to ventilation requirements, however, there is first a discussion of the various terms that are used for expressing the quantity of ventilation being provided. This is followed by a discussion of the actual guidelines and standards that exist that pertain to IAQ in nonindustrial buildings, both from the ventilation and contaminant approaches.

TERMS AND UNITS FOR EXPRESSING VENTILATION RATES

The evaluation of the performance of a ventilation system is primarily concerned with the quantity of outdoor air delivered to the building occupants. This quantity of outdoor air can be expressed in several different units. Each of these differing evaluation units will therefore have its own criteria for comparison purposes. These evaluation yardsticks can be expressed in units which include the absolute measurements of the outdoor air quantity, in cubic feet per minute (cfm)

Table 4.1.   Evaluation Units for Quantifying Ventilation Quantities

CFM of outdoor air
Percentage of outdoor air in the supply air
CFM of outdoor air per person
CFM of outdoor air per square foot of building area
Air changes per hour of ventilation

or liters per second of OA, as well as units which compute this quantity of outdoor air in relation to such parameters as the total volume of delivered supply air, the number of people present, the floor area, or the building volume. Specifically, the evaluation criteria for assessing the ventilation quantities are listed in Table 4.1.

For each of the five different terms that can be used to quantify a ventilation rate in a building, or portion of that building, there is a relationship that will permit all of these terms to be calculated from each other if the details of the building and its population are known. The significance of each of these different ways of expressing the quantity of outdoor air can be summarized as follows.

CFM of Outdoor Air

The term "cfm of outdoor air" is concerned with the absolute quantity of outdoor air. The major importance of this terminology is that the original design specifications for the building will typically be expressed in these units and therefore will appear in the mechanical drawings and specifications for a given building. The location in the drawings to check is in a table entitled "Schedule of Equipment" which lists the details of the equipment and their designed capabilities. This schedule will typically list the minimum outdoor air quantity in cfm, as well as the total supply air quantity. This value, therefore, can be considered as the "intended" goal of the ventilation system. This information is important for the evaluation because, as described in Chapter 5, there needs to be an assessment performed to determine if the system is in fact performing as intended. In evaluating the ventilation system and the adequacy of the delivered quantities of outdoor air, three basic scenarios need to be considered; these include:

1. The system is performing as intended; however, this volume of outdoor air is inadequate for the current use or density.
2. The design quantities would be adequate if achieved; however, the system is not performing as intended.
3. The design quantities would be inadequate even if they were being achieved, and they are not.

Percentage of Outdoor Air in the Supply Air

The use of these units reflect the fact that the supply air is made up of a mixture of both the outdoor air for ventilation and air that is recirculated from the building which are then combined to achieve a total volume needed to facilitate thermal regulation. Buildings are usually operated with a minimum setting of outdoor air at 15 to 20% of the total supply air volume. Since the amount of supply air varies from building to building, this determination is useful as a rough or, ball-park, evaluation criterion.

In actual practice, the percentage of outdoor air can vary from 0% all the way up to 100%, depending on the design, installation, and operation of the equipment. For a given HVAC system, this percentage can change in response to changes in the outdoor air temperature. When the building is occupied, there must always be some minimum percentage of outdoor air in the supply air, in accordance with the requirements of building codes. When a building evaluation is being performed, this potential variation needs to be taken into consideration. That is, before evaluation measurements are performed, the operating characteristics of the ventilation system need to be understood. Once this is accomplished, the evaluation measurements can proceed, knowing their relationship to the potential minimum conditions. In fact, it is recommended that the evaluation include a quantification of the amount of ventilation being provided during these minimum conditions. Situations where the outdoor quantity is down near 0% can arise for a number of reasons, all of which involve both the position of the outdoor air (OA) dampers and the pressure differential across the mixing box. For more information on these two topics, refer to Chapters 2 and 8.

The 100% outdoor air situation can occur for two basic reasons. One situation involves the continuous delivery of 100% outdoor air because there are activities occurring in the building, such as in the case of a chemical laboratory, which preclude the recirculation of air from those spaces. The second situation involves the intermittent delivery of 100% outdoor air because the HVAC includes full economizer capability which permits the use of "free cooling" when the outdoor air temperature is equal to the desired supply air temperature (less the temperature increase caused by the heat given off by the fan motor). This type of installation is highly desirable, both from an IAQ and energy conservation standpoint.

CFM of Outdoor Air per Person

One approach for quantifying ventilation rates is to use a calculation where the absolute quantity of outdoor air being delivered is divided by the number of people present or potentially present. Expressing the ventilation rate in units of cfm of outdoor air per person is a very important evaluation criterion because the *ASHRAE Standard 62 Ventilation for Acceptable Indoor Air Quality*[1] lists minimum recommended ventilation rates expressed in these units. For offices, this voluntary consensus standard specifies a minimum outdoor air requirement of the delivery of 20 cfm per person and is based on an estimated maximum occupancy of 7 people per 1000 ft^2 (or 100 m^2), where the area represents net occupiable space. Net occupiable space refers to the floor area portion of a building allocated for the office spaces and does not include utility areas or spaces used as mechanical rooms. Other spaces, such as smoking lounges, are listed as having an outdoor air requirement of 60 cfm per person as exhaust. For smoking lounges, the estimated maximum occupancy is 70 people per 1000 ft^2 (or 100 m^2). For classrooms, the minimum outdoor air requirement is 15 cfm per person and the estimated maximum density is 50 people per 1000 ft^2.

Another very important consideration with respect to ventilation rates expressed in the units of cfm per person is the fact that the carbon dioxide (CO_2) evaluation technique can potentially yield results in exactly these units, thus permitting a direct comparison with the requirements of *ASHRAE Standard 62-1989*.[1] The specifics of this evaluation technique are presented in Chapter 6.

CFM of Outdoor Air per Square Foot

Expressing the ventilation rate in terms of outdoor air per area of space is a form taken by several corporate standards for their buildings and the ventilation requirements of the California Title 24 regulations. One such corporate standard, for instance, requires 0.15 cfm of outdoor air per square foot of occupiable area. For example, if the total supply air rate is 1.0 cfm per ft^2, a typical minimum value, then the outdoor air quantity would need to be equal to 15% of the supply air. The importance of this ventilation rate evaluation criterion is its recognition of the fact that people are not the only interior source of air contaminants in the building, and that the furnishing and building components also have the potential to be sources of air contaminants. When using this term, the evaluator needs to be precise as to the definition of the floor area to be considered in this calculation. Typically, this ventilation rate term uses the units of net occupiable area or net rentable area and not the gross area of the building. Applying this ventilation rate therefore requires a detailed knowledge of how the different areas of the space being evaluated are designated and intended for use.

Other locations where the activities, and not the number of people, are the predominant source of air contaminants will have their minimum outdoor air requirements specified in the units of cfm per ft^2. For example, ASHRAE Standard 62-1989[1] calls for 0.5 cfm of OA per ft^2 in printing and duplicating rooms.

Air Changes per Hour of Ventilation

Expressing the ventilation rate in terms of the quantity of outdoor air divided by the building volume is another approach that is frequently used. Dividing a volumetric flow rate term (like cfm) by a volume term yields a value that can be expressed in the units of air changes per hour (ACH). This approach is important because these units express the rate at which the ventilation system actually dilutes and removes the air contaminants present in the space under evaluation. When expressing a quantity in terms of ACH, it is important to specify that it is either a ventilation rate, in the units of air changes per hour of outdoor air (ACH of OA), or that it is the total supply air quantity which is also expressed in the units of air changes per hour. Total supply air rates are typically in the range of 5 to 7 ACH (total air), while minimum ventilation rates are typically around 1.0 ACH of OA. These typical values correspond to a range of 14 to 20% outdoor air in the supply air.

Another significant fact about expressing the ventilation rate in the units of ACH of OA is that the results from tracer determinations of effective ventilation rates are also presented in these units. One description of this testing technique is presented in *ASTM E741-83, Standard Test Method for Determining Air Leakage Rate by Tracer Dilution.*[2]

The challenge for converting between ventilation rates in ACH and the other units is that the effective volume of the space under evaluation must also be known. This effective volume is the volume that the tracer, and the air in the space in general, has been free to move through. The effective volume will be less than the gross interior volume by the amount equal to the volume of the interior furnishings (i.e., file cabinets, desks, etc.), the volume of the people, and the volume of the stagnant areas, such as near corners and in closets. Although the effective volume can be determined by constant emission tracer testing, this procedure is difficult and time consuming and frequently cannot be justified in economic terms.

Experience has indicated that the effective volume of furnished and occupied buildings can be estimated at 85% of the gross interior volume, furnished but unoccupied buildings can be estimated at 90% of the gross interior volume, and buildings that are both unfurnished and unoccupied can be estimated at 95% of the gross interior volume.

Another important aspect to remember with respect to the usage of the "air changes per hour" term is that it refers to the volumetric equivalent of the effective volume; thus, after 1 hour at 1.0 ACH, the entire contents of the space have not necessarily been replaced. The relationship between the volume of air actually replaced and the ventilation rate in ACH is a function of both the mixing characteristics and the uniformity of airflow in that space. The resulting three basic categories of ventilation type are summarized in Table 4.2.

For the typical well-mixed ventilation by dilution situation, the amount of air actually replaced after 1 hour at 1.0 ACH will theoretically be 63.2%. For ventilation by displacement, or *plug-flow*, the amount of air replaced after 1 hour at 1.0 ACH will theoretically be 100%. For the short-circuiting case, the amount of air replaced will be less than the 63% in the well-mixed case; how much less will depend on the magnitude and specifics of the short-circuiting. With short-circuiting, the ventilated space becomes not just one zone but several, with differing zones receiving varying replacement rates.

Table 4.2. Ventilation Type Categories

Ventilation Type	Mixing Characteristics	Uniformity of Flow
Dilution	Well mixed	Uniform
Displacement	Not well Mixed	Uniform (linear)
Short-Circuiting	Not well Mixed	Non-uniform

These differences in ventilation type correspond to variations in how the air moves through the occupied space and is referred to as the ventilation efficiency. This topic is discussed in more detail in Chapter 7 and ASHRAE Standard 62-1989.

TYPICAL RELATIONSHIPS AMONG THE DIFFERING VENTILATION RATE UNITS

For typical spaces with ceiling heights of 12 ft overall, based on 10 ft below the suspended ceilings and 2 ft above for the return plenum, the following relationships exist among the various units for expressing ventilation rates.

Case 1: Typical situation (base case)

Assumptions and dimensions:
   Area = 10,000 ft^2 (80 × 125 × 12 ft with suspended ceiling at 10 ft)
   Density = 7 people per 1000 ft^2
   Supply air quantity delivered = 10,000 cfm
   Outdoor air quantity delivered = 1400 cfm
Ventilation rates:
   14% outdoor air in supply air
   20 cfm of OA/person
   0.14 cfm of OA/ft^2
Gross interior volume = 120,000 ft^3
Estimated effective volume = 102,000 ft^3 (85% of gross)

Ventilation rate = 0.82 ACH of OA

Case 2: Building with extra high ceiling plenum

Assumptions and Dimensions:
   Area = 10,000 ft^2 (80 × 125 × 18 ft with suspended ceiling at 10 ft)
   Density = 7 people per 1000 square feet
   Supply air quantity delivered = 10,000 cfm
   Outdoor air quantity delivered = 1400 cfm

Ventilation rates:
   14% Outdoor air in supply air
   20 cfm of OA/person,
   0.14 cfm of OA/ft^2
Gross interior volume = 180,000 ft^3
Estimated effective volume = 153,000 ft^3 (85% of gross)

Ventilation rate = 0.55 ACH of OA

Increasing the overall floor-to-ceiling height increases the volume of air involved, thus reducing the ventilation rate; in this case, from 0.82 to 0.55 ACH. However, the ventilation rate on a per person or square foot basis remains the same. The trade-off is that, although there is a larger volume of air for initial dilution of air contaminants, these contaminants will persist for a longer interval of time in the Case 2 situation. The half-life of air contaminants at 0.82 ACH is 51 min, as compared with 1 hour 16 min at a ventilation rate of 0.55. At these rates, for the situation where the air contaminant was no longer being released into the space, the time required to reduce the concentration of an air contaminant by 90% would be 2 hours 48 min at 0.82 ACH, while it would take 4 hours 11 min at 0.55 ACH.

Case 3: VAV system at part-load (throttled) condition

Assumptions and dimensions:
   Area = 10,000 ft^2 (80 × 125 × 12 ft with suspended ceiling at 10 ft)
   Density = 7 people per 1000 square feet
   Supply Air Quantity Delivered = 6500 cfm
   Outdoor Air Quantity Delivered = 910 cfm
Ventilation Rates:
   14% Outdoor air in supply air
   13 cfm of OA/person
   0.09 cfm of OA/ft^2
Gross Interior Volume = 120,000 ft^3
Estimated Effective Volume = 102,000 ft^3 (85% of gross)

Ventilation rate = 0.54 ACH of OA

In Case 3, the total quantity of supply air has been reduced to 65% of the design value, from 10,000 cfm down to 6500 cfm. The outdoor air, while remaining at 14% of the supply air, has similarly been reduced to 910 cfm. The number of people present has remained the same, but this measure of the ventilation rate has dropped to 13 cfm outdoor air per person. In addition, the ventilation rate based on the volume has also dropped down to 0.535 ACH of OA.

VENTILATION RATES AND THE HVAC SYSTEM

In addition to changes in relative volumes and population density, the HVAC system of course has a direct effect on the ventilation rate being provided. With respect to the HVAC system, its design, installation, operation, and maintenance can all contribute to the resulting ventilation rates. The factors to consider for each of these aspects of the HVAC system or systems are discussed in detail in Chapter 5.

Summarizing the discussion of the various terms available for quantifying ventilation rates, the most basic approach is to use their absolute quantity, expressed as the cfm of outdoor air. This quantity of outdoor air is intimately related to the design of certain aspects of the HVAC equipment. Both the specifications for the capacity of the cooling and heating coils are based on the requirements of summer and the winter design conditions, respectively. That is, for example, if the winter design condition is for outdoor air temperatures of 0°F and the design minimum outdoor air quantity is 10,000 cfm, then the heating coils will be sized to be able to heat this quantity of air up to the required delivery temperature for supply air. This information will be included in the mechanical drawings for the building and a review of this information should be included as part of the evaluation process. The other information to be gleaned from these tables is the quantity of total supply air which is intended to provide thermal comfort based on anticipated thermal loads in the building.

STANDARDS, GUIDELINES, AND REGULATIONS

As stated at the beginning of this chapter, standards, guidelines, or regulations that pertain to the indoor environment can be structured to specify either minimum quantities of ventilation air or maximum concentrations of air contaminants that are allowable, either as concentrations in the air or as emission rates from building materials. Ventilation standards include those by ASHRAE, HUD, model building codes, and the APHA Model Housing Code.

ASHRAE Standard 62-1989

ASHRAE Standard 62-1989, Ventilation for Acceptable Indoor Air Quality[1] lists minimum outdoor air requirements which are expected to be deemed capable of providing an acceptable level of IAQ. This standard was developed by an interdisciplinary committee and it reflects a consensus opinion that attempts to balance the requirements of acceptable IAQ and efficiency in energy consumption. Acceptable IAQ is defined by ASHRAE as "air in which there are no known contaminants at harmful concentrations as determined by cognizant authorities and with which a substantial majority (80% or more) of the people exposed do not express dissatisfaction" upon survey. This standard therefore recognizes the dynamic relationship between ventilation rates and air contaminants, and that merely addressing one cannot completely define the requirements for achieving good IAQ. ASHRAE also recognizes that this standard does not, and cannot, assure that no adverse health effects will occur.

ASHRAE Standard 62-1989 also needs to be recognized in the steps it takes in going beyond just specifying the requirements for minimum outdoor air quantities. There are several other requirements of the standard that are also very significant in their role in helping to achieve good IAQ. These additional requirements recognize the roles of the design, installation, operation, and maintenance of the HVAC system in both providing adequate quantities of outdoor air and minimizing or eliminating the presence of air contaminants. For instance, in Section 5, "Systems and Equipment," the following requirements are enumerated.

Section 5.1: "When mechanical ventilation is used, provision for air flow measurement should be included."

Section 5.2: "Ventilating systems shall be designed and installed so the ventilation air is supplied throughout the occupied zone. The design documentation shall clearly state the assumptions that were made in the design with respect to ventilation rates and air distribution."

Section 5.4: "When the supply of air is reduced during times the space is occupied (e.g., in variable-air-volume systems), provision shall be made to maintain acceptable indoor air quality throughout the occupied zone."

Section 5.5: "Ventilating systems should be designed to prevent reentrainment of exhaust contaminants. Contaminants from sources such as cooling towers, sanitary vents, vehicular exhaust from parking garages, loading docks, and street traffic should be avoided."

Section 5.6: "Ventilating ducts and plenums shall be constructed and maintained to minimize the opportunity for growth and dissemination of microorganisms through the ventilation system."

Section 5.7: "Contaminants from stationary local sources within the space shall be controlled by collection and removal as close to the sources as practicable."

Section 5.10: "When compliance with this section does not provide adequate control of gaseous contaminants, methods based on sorption or other scientifically proven technology shall be used."

Section 5.12: "Microbial contamination in buildings is often a function of moisture incursion from sources such as stagnant water in HVAC air distribution systems and cooling towers. Air-handling unit condensate pans shall be designed for self-drainage to preclude the build-up of microbial slime. Provision shall be made for periodic in-situ cleaning of cooling coils and condensate pans. Air-handling and fan-coil units shall be easily accessible for inspection and preventive maintenance."

Other significant requirements of this standard are also included in Section 6, "Procedures" which states that, "Indoor air quality is a function of many parameters including outdoor air quality, the design of enclosed spaces, the design of the ventilation system, the way this system is operated and maintained, and the presence of sources of contaminants and the strength of these sources." Included in this section are the following requirements.

"The quality of the outdoor air needs to meet the National Primary Ambient Air Quality Standards for Outdoor Air as set by the U.S. Environmental Protection Agency, or be treated to control the offending contaminants."

"Design documentation shall clearly state which assumptions were used in the design so that the limits of the system in removing contaminants can be evaluated by others before the system is operated in a different mode or before new sources are introduced into the space."

As an additional point, Section 6.1.3.3 can be interpreted to require that the delivery of the specified quantities of outside air must be to the breathing zone of the building occupants.

ASHRAE Standard 62-1989 is published by the American Society of Heating, Refrigerating and Air-Conditioning Engineers, 1791 Tullie Circle, N.E., Atlanta, GA. Copies of the complete standard may be purchased from the Society.

HUD Ventilation Requirements

HUD ventilation requirements are incorporated into its minimum property standards for residences[3] which are part of its mortgage insurance and low rent public housing program and in construction requirements for manufactured housing.[4] Ventilation for the construction of manufactured housing can be met in two ways. The rule specifies that an area equivalent to not less than 8% of the floor area must be available for natural ventilation (windows or doors); or alternatively, a mechanical system must be capable of a ventilation rate of 2 ACH of OA. Bathrooms and toilet compartments require either 1.5 ft^2 of openable glazed area or a mechanical system capable of producing 5 ACH. The mechanical system must exhaust directly outside the house. It should be noted that since this requirement for mechanical ventilation is only for a given capability, it does not say anything about actual operational requirements.

The rule for manufactured housing construction specifies the venting of combustion appliances and requires purchasers to be presented with options to improve overall ventilation. HUD is now considering ways to improve this standard.

State Legislation

While there are consistent nationwide, federal standards that have been developed specifically for non-industrial environments, several states have promulgated regulations in this area. Legislation passed by the states can be divided into the following four categories:

1. Indoor Air Programs/Coordinating Committees
   Legislation directs a state agency or agencies, or an appointed commission or board to either develop an indoor air program and make recommendations for implementation of such a program, or recommends that the body do research to better characterize the indoor air problems in the particulate state.

2. Ventilation Standards
   This legislation mandates compliance with standards already in existence (ASHRAE), or establishes new standards and extra ventilation in particular places.

3. Pollutant Specific Standards

   This legislation involves development of numerical standards for particular pollutants. In some cases, these include fairly standard pollutants, like asbestos, radon, or formaldehyde, but in others (i.e. New York) the legislation takes a health advisory approach, recommending research and development of standards for a number of different pollutants.

4. Smoking

   By far the most common legislation from the states involves restricting smoking to designated places, or restricting it completely in other places.

Other specific examples include the State of California where Subchapter 3 (Sections 120 through 129) of the new Title 24 Ventilation Regulations specify a minimum ventilation rate, for different types of use. All uses not specifically listed require a minimum of 0.15 cfm of OA per ft^2 of conditioned floor area. These Requirements for Ventilation (Section 121) apply to all enclosed spaces in a building that are normally used by humans.

Ventilation Requirements in Model Codes

Building codes identify design and construction specifications for buildings. The primary building codes that are in use in the U.S. include those written by the Building Officials and Code Administrators International (BOCA), the Southern Building Code Congress International (SBCCI), the Council of American Building Officials (CABO), and the American Public Health Association (APHA) model code.

These codes are updated periodically to reflect new knowledge and incorporate standards developed by other organizations. State and local governments can either adopt these codes in their entirety or revise them as needed. Sufficient time has not yet elapsed for these organizations to consider incorporating provisions of the new *ASHRAE Standard 62-1989*. This step is however being considered.

The CABO code, for one- and two-family dwellings, specifies ventilation requirements for habitable rooms in terms of openable windows or mechanical ventilation systems that can provide 2 ACH of OA. For bathrooms, toilet compartments, and similar rooms, the ventilation requirements can again be achieved by operable windows or a mechanical ventilation system that can provide 5 ACH of OA.

The APHA model code[5] also specifies ventilation air requirements. In the model code, ventilation requirements can be met by openable windows or mechanical ventilation systems. The code also requires HVAC units which are integral to the structure to be operated continuously; if the unit becomes inoperable, alternate provisions for ventilation are to be provided.

The code also contains a provision that requires ventilation, either natural or mechanical, to provide acceptable IAQ in every habitable room at all times when occupied. In addition, bathroom and kitchen exhaust air cannot be recirculated.

The model code does not specify numerical limits for indoor air contaminants. Instead, it defines acceptable IAQ as indoor air in which there are no known concentrations which are in excess of those which have been established by the Director of Health. This provision provides local jurisdictions with additional flexibility and control because the local health officer has the authority to declare a particular situation a health hazard and require remediation.

I have found that in the view of many building inspectors, for instance, as long as the ventilation system was designed to meet the ventilation requirements of the building code at the time the building permit was obtained, the building is considered in compliance. There are currently many forces at work to introduce more regulation in this area; it is not clear yet exactly what form this will take.

This discussion, however, would not be complete without mention of the use of carbon dioxide (CO_2) concentrations of the not-to-exceed variety as a guideline for ventilation rates. The Department of Public Health of the Commonwealth of Massachusetts has a Comfort Guideline that stipulates that the CO_2 concentration indoors should not exceed 800 ppm. While this is a reasonable approach, there are some problems with this tactic. The major shortcoming of using this technique is due to the fact that ventilation rate actually relates to the increase of the indoor concentration over the outdoor concentration, and the outdoor CO_2 concentration ranges from an average of 350 ppm down to 300 ppm and up to 450 ppm. Using the equation that relates the difference in indoor and outdoor CO_2 concentrations to an effective ventilation rate (see Equation 6.2), for the 800 ppm maximum indoor concentration, this variation in outdoor CO_2 concentration can correspond to a minimum ventilation rate of 30.3 cfm per person of outdoor air when the outdoor value is 450 ppm, down to a minimum ventilation rate of 21.2 cfm per person of outdoor air when the outdoor value is 300 ppm. For the average outdoor reading of 350 ppm, the minimum required ventilation rate is 23.6 cfm per person of outdoor air. In my opinion, therefore, the resulting arbitrariness of this approach may limit its widespread adoption. The same basic shortcoming applies to a not-to-exceed value of 1000 ppm of CO_2 as well.

Air Contaminant Standards

Air quality standards and guidelines specify maximum concentrations and exposure times for specific contaminants in indoor and outdoor environments. They are designed to protect specified classes of individuals from adverse health impacts. Different standards are designed to protect different classes of individuals, and to different degrees, depending on the nature and purpose of the standards.

There are two broad types of standards and guidelines: public health standards and occupational standards. Public health standards are designed to protect the general public and are therefore most applicable to residential, educational, commercial, and public building environments. Occupational standards, on the other hand, are applied to industrial work environments.

Public health standards are generally one to two orders of magnitude lower (more protective) than occupational standards for several reasons.

1. Occupational standards cover robust healthy adult workers, while public health standards are designed to protect all segments of the population, including potentially sensitive subgroups — the elderly, infants and children, pregnant women, and those with preexisting heart and lung diseases.
2. Occupational standards are for workers for limited periods of time (usually the 8-hour work day), while public health standards generally protect the public for continuous lifetime exposures (24 hours per day).
3. Occupational standards factor cost and technical feasibility into the recommended limits, while public health standards are often established without regard to cost or feasibility.

As stated before, *there are no federal standards in the U.S. that have been developed specifically for indoor air contaminants in nonoccupational environments.* There are related existing standards and guidelines that can be used for some comparisons, but most of these are not directly applicable to indoor air quality problems. For instance, in the category of public health standards and guidelines, there are the National Ambient Air Quality Standards.[6] These standards, which are presented in Table 4.3, may not be directly applicable to indoor air environments because of differences in averaging times and the lack of a clearly defined enforcement mechanism.

The other existing category of standards and guidelines for air contaminants are the occupational ones, consisting of the American Conference of Governmental Industrial Hygienists (ACGIH) Guidelines and the Occupational Safety and Health Administration (OSHA) Standards. The ACGIH regularly reviews the best available information from industrial, epidemiological, and animal studies to develop new limits and revise old ones.

ACGIH first published its limits, *Threshold Limit Values (TLVs),* in 1968 and they are updated annually.[7] The TLVs are concentration limits and conditions to

Table 4.3. National Primary Ambient Air Quality Standards for Outdoor Air as Set by the U.S. EPA

	Long term			Short term		
	Concentration			Concentration		
Contaminant	$\mu g/m^3$	ppm	Averaging	$\mu g/m^3$	ppm	Averaging
Sulfur dioxide	80.0	0.03	1 year	365	0.14	24 hours
Total particulate	75.0	—	1 year	260	—	24 hours
Carbon monoxide				40,000	35	1 hour
Carbon monoxide				10,000	9	8 hours
Oxidants (ozone)				235	0.12	1 hour
Nitrogen dioxide	100.0	0.055	1 year			
Lead	1.5	—	3 months			

From "National Primary and Secondary Ambient Air Quality Standards," *Code of Federal Regulations,* Title 40 Part 50 (40CFR50).

which it is reported that nearly all workers may be repeatedly exposed to day after day throughout their working lifetime without adverse effects. However, because of the individual variation in response to chemical exposure, the ACGIH recognizes that a small percentage of workers may experience discomfort at levels below the TLVs, and an even smaller percentage may experience more serious effects such as aggravation of a preexisting condition or the development of an occupational illness at levels below the TLVs.

OSHA is responsible for protecting workers from unsafe and unhealthy work environments. Standards for the workplace are promulgated by the Secretary of Labor under the Occupational Safety and Health Act of 1970.[8] The ACGIH TLVs have been the basis for U.S. standards when OSHA has twice (1968 and 1989) adopted essentially the entire list of TLVs for chemical substances as enforceable limits. The exposure standards for air contaminants are intended to be set at levels that will protect a worker from "material impairment of health or functional capacity" even if the worker is exposed for 8 hours per day for an entire working lifetime. The occupational health standards are based on a healthy adult worker; they do not take into account the variability of the general population.

Unfortunately, there have been many building investigations where OSHA standards for air contaminants were used as evaluation criteria. Not only was this approach a waste of money and time, but when the results were reported as "all well below OSHA Standards and therefore there was nothing wrong with the air quality," a grave disservice was done to the employees. Although there are those in the AIHA (American Industrial Hygiene Association) who believe that occupational threshold limit values (TLVs) for indoor contaminants are applicable to the non-industrial environment such as offices, schools, stores, etc., this is at odds with my own experience and those of IAQ professionals whose opinions I respect. Hal Levin,[9] in the *Indoor Air Quality Update,* makes the following points when discussing the rationale for applying factors to reduce TLVs when applied to indoor air exposures. This listing can also be used to explain why TLVs are inappropriate to use in an non-industrial environment.

1. In non-industrial settings, people are exposed to many chemical substances simultaneously and serially; whereas, it is assumed that in the industrial workplace, there is a very limited number of hazardous substances requiring control.
2. The industrial setting is regulated and, presumably, monitored to ensure compliance with the relevant standards. Monitoring is almost never done in indoor air settings. Therefore, elevated levels could occur without the knowledge of the exposed individuals. Excursions (short-term elevations of air concentrations) can occur as a result of using some product or performing some task. Some such excursions have been shown to constitute most of an individual's total exposure to the relevant substances.
3. In the industrial setting, devices such as fume hoods are installed and operated to control air concentrations of hazardous substances where they are known to occur. Workers wear protective clothing and breathing apparatus when necessary. Also, medical surveillance provides detection at early stages of adverse

effects. None of these protections are present, at least not routinely, in the indoor air setting.

4. The industrial workplace is primarily occupied by healthy adults, in many cases mostly males; whereas, the general indoor environment is occupied by the entire population spectrum, including the elderly, infants, infirm, pregnant women, and populations susceptible due to allergies, sensitivities, or other predispositions. In addition, new workers in an industrial setting who exhibit a sensitivity to their exposures tend not to remain in that occupation. Therefore, the industrial population exposed to a particular contaminant tends to be composed of individuals with lower sensitivity to the particular contaminant or contaminants.

5. TLVs (and other occupational standards) are based on an 8-hour work-day, 40-hour work-week exposure. This assumes less than one-fourth the number of hours of exposure that might occur for persons staying in a single environment or otherwise exposed to a substance continuously throughout the 168-hour week. The studies upon which the standards are supposedly based generally involve the work-week exposure. Further, the work-week exposure allows time for recovery between exposures — 16 hours between shifts and 64 hours on the weekend.

6. ACGIH states in Appendix C, "TLVs for Mixtures," that the TLVs are to be applied in a manner that considers, in an additive way, all exposures affecting the same target organ system. "When two or more hazardous substances, which act upon the same organ system, are present, their combined effect, rather than that of either individually, should be given primary consideration. In the absence of information to the contrary, the effects of the different hazards should be considered as additive."

Indoor air is a complex mixture of hundreds of substances, dozens of which are likely to act upon the same target organ system. Therefore, if a person is exposed to six substances that are known to affect the central nervous system, for example, the concentrations of each substance divided by their respective TLVs are added; if the sum exceeds 1, then it is considered that the TLV for the mixture has been exceeded.

The additive approach, however, may not provide adequate protection since it is possible that there are synergistic effects (the combined effects are greater than the additive effects). For example, the increased cancer risk from occupational asbestos exposure alone is approximately fivefold and from tobacco smoking alone is approximately 11-fold. The increased risk from the combined exposure is about 52-fold, closer to the multiple of the risks from the separate exposures than to their sum. Using the additive approach, the calculated increased risk is 16-fold, less than one fourth the known risk. Note that this example deals with carcinogenicity, and the TLVs treat carcinogens differently from the other endpoints of narcosis, irritation, and impairment of health.

These points were raised to explain the rationale for applying factors to reduce TLVs when applied to non-industrial air exposures; however, the question that has not been addressed is, by what factor — 10, 100, or even 1000 — would the TLV need to be reduced to provide the appropriate amount of protection? There has been no systematic investigation into this issue that I know of.

In some of the ongoing discussion on this issue, a case has been made for the fact that the so-called "Threshold Limit Values" do not in fact represent thresholds. In one study, "But They Are Not Thresholds: A Critical Analysis of the Documentation of Threshold Limit Values," it was determined that when exposures were at or below the TLV, *only a minority of studies showed no adverse effects (11 instances out of 158).*[10] The authors conclude that "TLVs for chemical substances are a compromise between health-based considerations and strictly practical industrial considerations, with the balance seeming to strongly favor the latter. In other words, most TLVs may represent guides of levels which have been achieved *but they are not thresholds*. We therefore regard the definition of ACGIH TLVs as incorrect and the term "threshold" in the name of the limits as singularly inappropriate."

The significance of this finding is that the suggested approach of dividing industrial occupational standards by an arbitrarily selected protection factor (sometimes 10 to 1000) to create a benchmark for non-industrial environments is on shaky ground. This practice is not recommended for the additional reason that these standards were not derived to protect the general population, and the application of protection factors can result in significant inconsistencies among contaminants. The resulting numbers may not provide sufficient protection, or they may be unduly restrictive.

Setting standards for IAQ contaminants is in its infancy. One perspective on where the regulation of indoor air might be going in the future was offered by Seifert of the West German Institute for Water, Soil and Air Hygiene at the *INDOOR AIR '90 Conference.*[11] He commented that setting standards for indoor air would not be an appropriate way to reduce indoor air pollution, in contrast to the situation for outdoor air and the air at industrial workplaces. Rather, he suggested that guideline values for pollutant concentrations could be utilized. Guideline values could be set at two levels: the first level defines the hazard, while the second identifies a target concentration for the future. He also pointed out that for individual substances, sufficient toxicological information may be available to derive such guideline values, but mixtures are not as easily evaluated.

As an example, he suggested a target guideline value for total volatile organic compounds (TVOCs) of 0.3 mg/m^3. His definition of TVOC refers to the sum of individual VOCs separated and quantified by a gas chromatographic technique. To obtain the sum, the VOCs belonging to particular chemical classes are ranked according to their measured concentration, as presented in Table 4.4.

Although Seifert stresses that the guideline is based on his own judgement, and *not* on toxicological data, he feels that they reflect what could be achieved based on reports in the scientific literature. The values given are for long-term exposure. After construction or renovation, for 1 and 6 weeks, respectively, he proposes that concentrations 50 and 10 times higher be acceptable. In a related plenary lecture at *INDOOR AIR '90,* Mølhave[12] of Denmark summarized field investigations and controlled experiments on the relation between low levels of indoor air pollution with VOCs and human health. He also suggests a tentative guideline for VOCs indoors. His findings are summarized in Table 4.5.

Table 4.4.   Proposed Target Guideline for TVOC in Indoor Air

Chemical Class of VOC	Concentration $\mu g/m^3$
Alkanes	100
Aromatic hydrocarbons	50
Terpenes	30
Halocarbons	30
Esters	20
Aldehydes and ketones (excluding formaldehyde)	20
Other	50
Target guideline value (sum of VOCs)	300

Note: The concentration of an individual compound should neither exceed 50% of the concentration allotted to its class nor 10% of the TVOC concentration.

From Seifert, B. 1990. "Regulating Indoor Air." *Indoor Air '90: Proc. 5th Int. Conf. Indoor Air Quality and Climate.* Toronto, Canada. 5:35–49. With permission.

Table 4.5.   Tentative Dose-Response Relationship for Discomfort Resulting from Exposure to Solvent-Like Volatile Organic Compounds

Total Concentration mg/m^3	Irritation and Discomfort	Exposure
<0.20	No irritation or discomfort	The comfort range
0.20 – 3	Irritation and discomfort possible if other exposures interact	The multifactorial exposure range
3.0 – 25	Exposure effect and probable headache possible if other exposures interact	The discomfort range
>25	Additional neurotoxic effects other than headache may occur	The toxic range

The overlap between Seifert's and Mølhave's recommendations could be interpreted to indicate that a consensus on recommendations for guideline values will evolve in the foreseeable future. Both of these men are members of the European Communities' Indoor Air Committee.

Source Emission Standards

Source emission standards are similar to air contaminant standards in that they attempt to address the air contaminant side of the IAQ equation. Air contaminant standards focus on the ambient concentrations, while the source emission standards address the selection of materials and products used in the space. A paper presented by Tucker[13] of the U.S. EPA describes the use of this approach as a regulatory mechanism for achieving good IAQ. Source control is important because without it, emissions can be too great to be adequately diluted and removed by ventilation, or removed by air-cleaning devices. This is especially true for sources that are close to the occupants, such as office furniture, furnishings, office supplies, and personal care products. An additional potential problem is that many sources release large quantities of pollutants over short periods of time; their emissions can be absorbed on many indoor surfaces (sinks) that subsequently reemit the pollutants and become significant sources themselves. The best currently available approach to evaluating indoor materials and products, is to test their emission rates and predict pollutant concentrations in the building where they are to be used. In fact, Tucker states that, "Emission testing and prediction of occupant exposures has recently become a key step in the design of some major office buildings." The state of Washington is already implementing a program for new state-owned office buildings which includes a requirement for systems office furniture (workstations) to have emission rates that will result in building air concentrations of less than the following values:

Formaldehyde ≤ 0.05 ppm (60 $\mu g/m^3$)
Total VOCs ≤ 0.50 mg/m^3 (500 $\mu g/m^3$)
Total particles ≤ 50 $\mu g/m^3$)

Whether the emissions from any source will lead to "acceptable" indoor concentrations and occupant exposures will depend on a number of considerations, namely: (1) emission rates, (2) the toxicity or irritation potential of substances emitted, (3) physical relationships between the source, the persons present, and the space they occupy, and (4) the sensitivity of the occupants.

In terms of the future development of emission standards as a viable control option for the achievement of good IAQ, an important step has already been accomplished. On September 28, 1990, ASTM[14] adopted the "Standard Guide for Determination of VOC Emissions in Environmental Chambers from Materials and Products." This standard has been designated as D5116-90. This guide will serve as the basis for more specific standards for emissions testing of various products and materials.

Thermal Comfort Standards

The achievement of thermally comfortable conditions, in addition to adequate ventilation and the absence of air contaminants, is a requirement for good indoor environment. Providing appropriate thermal environmental conditions is important in maximizing worker productivity and minimizing discomfort effects. There are many factors that influence the perception of comfort, temperature, and thermal acceptability. Important environmental factors include temperature, radiation, humidity, and air movement; personal factors include age, clothing, and activity level. *ASHRAE Standard 55-1981*[15] specifies thermal conditions which will be acceptable to 80% or more of the occupants in a building if dressed appropriately. According to this standard, the acceptable ranges of temperature and humidity during the summer and winter are summarized in Table 4.6.

Table 4.6.   Acceptable Temperature and Humidity Ranges

Relative Humidity	Winter	Summer
30%	68.5 – 76.0°F	74.0 – 80.0°F
40%	68.5 – 75.5°F	73.5 – 79.5°F
50%	68.5 – 74.5°F	73.0 – 79.0°F

Note: Applies for persons clothed in typical summer and winter clothing, at light, mainly sedentary activity. Humidities greater than 50% are considered unacceptable because of the potential for microbial growth.

From American Society of Heating, Refrigerating and Air Conditioning Engineers, Atlanta, GA 30329. Reprinted by permission from ASHRAE Standard 55-1981. Copyright 1981.

Lease Requirements

In the absence of federal, state, or local regulations pertaining to IAQ, individual tenants can make use of the building selection and lease negotiation processes as mechanisms for achieving good IAQ. While both the activities of the tenant and owners influence IAQ, including IAQ considerations in the lease will help spell out their responsibilities in preventing IAQ problems and establish procedures for responding to problems should they arise. Suggested lease clauses have been published by Prezant[16] that pertain to the following areas related to IAQ.

1. Quantities of outdoor air delivered to occupants
2. Air contaminant sources
3. Maintenance and operation
4. HVAC system capacities

There is clearly a lot of room for improvement in the use of leases for achieving good IAQ. In the leases I have reviewed in conjunction with IAQ investigations, either IAQ issues have not even been addressed or, if they have been addressed, the language is arbitrary. In one case, the lease indicated that the landlord would provide ventilation "equal in quality to those customarily provided by landlords in high quality office buildings in Boston."

On the other hand, however, as an example of the potential of this approach, is an Act of the Maine Legislature, to require state-leased buildings to meet certain air quality standards; it includes the following statements: "Application of minimum air ventilation standards. Beginning September 1, 1988, to apply the *ASHRAE Indoor Air Quality and Ventilation Standards* contained in the proposed revision, 1981R, July 15, 1986, as prepared by the American Society of Heating, Refrigeration and Air Conditioning Engineers, Inc. or more stringent standards to buildings occupied by state employees during normal working hours. These standards shall be applied ... to buildings for which the State enters into new leases following the date in this subsection."

Also, in the proposed *New Jersey Standards for Indoor Air Quality,* the following statement is included: "Preference in leasing or renting buildings shall be given to those buildings which meet the most recent practice consensus standards," *ASHRAE 62 and ASHRAE 55.*

SUMMARY AND CONCLUSIONS

Indoor air quality can be influenced by the development and implementation of ventilation standards, air contaminant standards, source emission standards, thermal comfort standards, and lease requirements.

The development of standards and guidelines for contaminant levels indoors and sources of contaminants is still in its formative stages. More research is needed before comprehensive standards can be developed that specifically address IAQ concerns.

Inadequate ventilation is one of the major causes of IAQ problems, and ventilation standards are an important method of control. Problems, however, do exist in relying on ventilation standards in building codes as the sole protector of the public health. Some underlying problems can be summarized as follows.

1. The current building stock has been constructed under different building codes with different requirements.
2. The current codes may be dominated by energy efficiency considerations, and code requirements may not be sufficient to provide adequate ventilation for indoor air quality purposes.
3. There is no guarantee that builders follow current code requirements or that they did so in the past.
4. Most jurisdictions do not have adequate inspection and enforcement capability to ensure that code requirements are followed.
5. Even if newer buildings are designed to meet the most current ventilation standards, improperly operated and poorly maintained systems will defeat the design goals and reduce ventilation rates.

In spite of the inadequacies in both ventilation and air quality contaminant standards, these two approaches can be used to facilitate the achievement of good IAQ in both old and new construction. New approaches, such as using performance standards for operating HVAC equipment, by proactive lease clauses or other methods, may provide additional measures to ensure adequate IAQ in commercial buildings.

REFERENCES

1. American Society of Heating, Refrigerating, and Air-Conditioning Engineers, Inc. 1989. *(ASHRAE) Standard 62-1989: Ventilation for Acceptable Indoor Air Quality.* Atlanta, GA.
2. American Society for Testing and Materials. 1983. "E 741-83, Standard Test Method for Determining Air Leakage Rate by Tracer Dilution." Philadelphia, PA.
3. U.S. Department of Housing and Urban Development (HUD). 1990. "Minimum Property Standards." *Code of Federal Regulations,* Title 24, Part 200, Subpart S, Sections 200.925 and 200.926.
4. U.S. Department of Housing and Urban Development (HUD). 1990. "Manufactured Home Construction and Safety Standards." *Code of Federal Regulations,* Title 24, Part 3280, Sections 3280.103 and 3280.710.
5. Mood, E., (Ed.). 1986. APHA-CDC *Recommended Minimum Housing Standards.* American Public Health Association, Committee on Housing and Health. Washington, D.C.
6. U.S. Environmental Protection Agency. "National Primary and Secondary Ambient Air Quality Standards," *Code of Federal Regulations,* Title 40 Part 50 (40CFR50) 1971.
7. ACGIH, 1988. *Threshold Limit Values and Biological Exposure Indices for 1988–1989.* Am. Conf. *Governmental Ind. Hygienists,* Cincinnati, OH.
8. U.S. Department of Labor Occupational Safety and Health Administration. OSHA Safety and Health Standards (29CFR1910). § 1910.1000.
9. Levin, H. 1990. *Indoor Air Quality Update,* 3:7 pp. 6–7.
10. Roach, S. A. and S. M. Rappaport. 1990. "But They Are Not Thresholds: A Critical Analysis of the Documentation of Threshold Limit Values." *Am. J. Ind. Med.,* 17:727–753.
11. Seifert, B. 1990. "Regulating Indoor Air." *Indoor Air '90: Proc. 5th Int. Conf. Indoor Air Quality and Climate.* Toronto, Canada. 5:35–49.
12. Mølhave, L. 1990. "Volatile Organic Compounds, Indoor Air Quality and Health." *Indoor Air '90: Proc. 5th Int. Conf. Indoor Air Quality and Climate.* Toronto, Canada. 5:15–33.
13. Tucker, G. 1990. Building With Low-Emitting Materials and Products: Where Do We Stand? *Indoor Air '90: Proc. 5th Int. Conf. Indoor Air Quality and Climate.* Toronto, Canada. 3:251–256.
14. ASTM. 1990. American Society for Testing and Materials. "Standard Guide for Determination of VOC Emissions in Environmental Chambers from Materials and Products."
15. American Society of Heating, Refrigerating, and Air-Conditioning Engineers, Inc. 1981. *(ASHRAE) Standard 55-1981: Thermal Environmental Conditions for Human Occupancy.* Atlanta, GA.
16. Prezant, B., D. Bearg, and W. Turner. 1990. "Tailoring Lease Specifications, Proposals, and Work Letter Agreements to Maximize Indoor Air Quality."*Indoor Air '90: Proc. 5th Int. Conf. Indoor Air Quality and Climate.* Toronto, Canada. 3:371–376.

CHAPTER 5

Evaluation of the Ventilation System

OVERVIEW

The investigation of ventilation systems can be summarized as requiring the items listed in Table 5.1. In terms of the organization of this book, the first three items listed in Table 5.1 are discussed in this chapter. The review of the occupant's complaints or concerns offer clues as to the condition of the building and its systems. The review of the design documentation provides a starting point for understanding the intent of the building and its systems. The measurement of the quantity of outdoor air (OA quantity) entering the HVAC system represents the bulk of the information presented in this chapter.

The measurement of the OA quantity delivered to the building occupants is presented and discussed in Chapter 6. Item 5, the determination of pathways of air movement into and through the building, is presented and discussed in Chapter 8. The final item in Table 5.1 is presented and discussed in Chapter 10.

Review of the Occupants' Complaints or Concerns

If the IAQ (indoor air quality) investigation is not being performed on a proactive basis, the first step in the evaluation process is to identify the historic issues of concern and complaints of the occupants of the building. This information helps to establish the scope of the investigation and represents clues as to what to look at first. The challenge in performing IAQ evaluations is the conflicting need to be both very thorough and very time efficient. While in some investigations every aspect of a building's system needs to be examined, in other studies

Table 5.1.   Requirements of Ventilation System Evaluations

A review of the occupant's complaints or concerns
A review of the design documentation
The measurement of the outdoor air quantity entering the HVAC System
The measurement of delivered quantities of outdoor air
A determination of air pathways through the building
Inspection for sources of air contaminants

some problems can be identified with just one walk-through. Occupant complaints represent clues as to where to focus the investigation. The complaints can be organized according to their timing, location, and type. The occupants are important sources of information about the functioning of the HVAC system due to the amount of time these people have spent in the building compared to the amount of time typically available for the investigator. In addition to reviewing the logs of complaints to the extent that they exist, scheduling time to talk to the individuals experiencing the discomfort, ill health, or fears will be both therapeutic for them and will permit the investigator an opportunity to ask questions and thereby obtain information typically not included in the logs.

With respect to the time of occurrence of complaints are problems being reported at certain times of the day or week, or in certain months or seasons of the year. Attempts should be made to determine if the pattern of complaints corresponds to any particular pattern of HVAC system operation, maintenance, or other observable changes or activities.

Another potential variable with the complaints are their distribution by location. Are the complaints associated with certain parts of the building? Are complaints concentrated in areas served by a single thermostat, zone, or air handling unit? Sometimes, complaints may be confined to certain floors or areas of a building, which will help focus the emphasis of the investigation.

With respect to the nature and type of the complaints themselves, are they similar or identical? It is important to be able to interpret the significance of the complaints; while the occupants will not intentionally attempt to confuse, their perceptions can be misleading. In one building, many people complained that the air was "too dry." This building however, was in Seattle, WA and the measurements determined that low humidity was not a problem. The reality was that the people were being exposed to elevated levels of particulates and VOCs (volatile organic compounds) and this combination was irritating to the eyes, which the people then perceived as being "too dry."

Indoor environments that are "too dry" (less than 30% relative humidities) will typically lead to complaints from contact lens wearers who will verbalize their discomfort and difficulty in being able to keep their lenses in all day.

Another example of the contrast between perceptions and reality was provided in a multidisciplinary evaluation of all aspects of the indoor environment. When asked their perception of noise levels, the typical response was that it was too noisy. The objective measurements, however, indicated that the office space had very low noise levels. The perception was based on the lack of acoustical privacy; voices carried. In this situation, by increasing the background levels of white noise, acoustical privacy was increased and people were happier.

While people's ability to detect odors exceeds that of most analytical equipment, their ability to explain and describe these odors is typically not as refined.

Another frequent complaint relates to the ease of transmission of colds and flu within the office. While this situation is indicative of low ventilation rates, due to the fact that viruses will persist in the air for longer intervals and at higher concentrations, it should be remembered that colds and flu are most frequently transmitted between individuals via the nose-to-hand-to-paper-to-hand-to-nose route.

Another perception-vs-reality complaint to be aware of are complaints of "paper fleas." There is no such insect according to the Entomology Office of the Department of Environmental Health and Safety at Harvard University. They report that past experience in office areas has found that micropaper particles and dust debris have a tendency to irritate the skin when the area is neglected by housekeeping services. Regular vacuuming by housekeeping should keep the debris to a level where employees would not be affected. They go on to comment that the ventilation system in the room should be inspected for proper operation.

Another building with IAQ complaints involved an "infestation of insects"; at least that was what was assumed due to the presence of "insect bites" on many of the arms of the building's occupants, which failed to diminish after repeated fumigations against potential insect vectors. The cause was ultimately determined to be the combination of low humidity levels and fine pieces of fiberglass insulation causing the irritation. It was a building in a northern climate, without winter humidification, that had fiberglass insulation batts installed around the perimeter of the building in the return air plenum, above the suspended ceiling. Since this location is part of the distribution system of the HVAC system, the fiberglass not filtered out was distributed throughout the building. If the complaints suggest the presence of insects, first collect samples and obtain identification of the actual species, if any, that are present. That is, if insects are actually present in a building, it should be possible to confirm this by finding some of their dead bodies as evidence of their presence.

I was involved in a residential IAQ evaluation where insects were actually a major contributing factor. In this building, the tenant and visitors were experiencing asthma and other respiratory complaints. The inspection of the heating and cooling system revealed the presence of insect bodies and construction debris in the supply air ductwork below the floor registers. The identification of the insects collected indicated the presence of centipedes, earwigs, and an Anobiidae beetle. The centipedes were the most prevalent. The presence of the centipedes was significant because they are predaceous on insects and other small arthropods, which therefore indicated that other smaller insects had been also present in the past in order to have provided food for them. The survival and multiplication of centipedes is also dependent on the presence of moisture; thus, their presence constitutes evidence of moisture in the heating/cooling system at some time in the past. The presence of the earwigs, which are chiefly plant feeders or scavengers, and the Anobiidae beetle, which feed on plant material such as wood, were indicators of the presence of biomass in this system. The construction debris observed to be present included sawdust and wood shavings — the indicated source of food. The other significant fact in this investigation was the establishment of the relationship between the presence of the insects and the asthma symptoms. A review of the literature established that the large proteins from living things, including insects, can cause occupational asthma.[1] Specific examples in the literature include exposure to insects such as the locust,[2] river fly,[3] screw worm,[4] cockroach,[5] bee moth,[6] and butterflies.[7] In fact, the Massachusetts Special Legislative Commission on Indoor Air Pollution[8] reported that "insects and insect parts are considered potential allergens which can lead to considerable acute respiratory and other discomforts, and long-term exposures can cause chronic conditions."

Review of the Design Documentation

The design documentation provides the basis for determining the original intent of the specification of the HVAC system. The investigator should therefore attempt to review all available information, which hopefully should include both original construction drawings as well as "as built" drawings which can frequently differ from the original construction drawings. The ability to read and comprehend blueprints are, of course, a useful skill in this portion of the investigation. In fact, one source of courses on "Improving Indoor Air Quality in Non-industrial Buildings" has added a new course entitled "Buildings: Understanding Systems and Blueprint Reading" starting in November 1991.[9] Knowing which locations in the building are at the end of the supply air distribution system is a critical piece of information because, sometimes, these locations will receive less air than the locations closer to the supply air source, and therefore may be the locations most likely to be receiving inadequate quantities of supply air.

When reviewing the building's mechanical drawings, in addition to seeing the layout of the equipment and ductwork, a very useful source of information is the "Equipment Schedule." This table typically contains the rated capacities of the equipment, including total supply air flow, maximum and minimum outdoor air flow, as well as heating and cooling capacities. This information then serves as a benchmark for the rest of the investigation, especially in that it indirectly states the loads assumed for the basis of the design and the design criteria. Two useful calculations relate to the density of people in the building and the predicted supply air flow per square foot of occupied area of the building. In the current *ASHRAE Standard 62-1989*,[10] the estimated maximum occupancy in various office spaces

Table 5.2. Comparison of Estimated Office Occupant Densities

Source/Location	Occupancy Density	
	ft²/person	people/1000ft²
ASHRAE 62-1989		
Office space	(142)	7
Reception areas	(17)	60
Telecommunications centers and data entry areas	(17)	60
Conference rooms	(20)	50
Carrier air conditioning		
Low	130	(7.7)
Average	110	(9.1)
High	80	(12.5)

is listed in Table 5.2. Also included in this table are low, average, and high values of occupant density from a Carrier pamphlet of 1975[11] for the building as a whole.

In Table 5.2, the values in parentheses are calculated from the published values from that source. With respect to supply air quantities, the Carrier pamphlet[11] lists 1.1 cubic feet per minute per square foot (cfm/ft^2) as an average value to controlling cooling loads for internal locations in office buildings. The low and high values are 0.8 and 1.8 cfm/ft^2 respectively.

THE MEASUREMENT OF THE OUTDOOR AIR QUANTITY

The determination of the OA quantity entering the HVAC system is an important starting point in a ventilation-based IAQ evaluation. The introduction of adequate quantities of outdoor air into the HVAC equipment is a necessary, although not sufficient condition for the maintenance of good IAQ. Not only must enough outdoor air be drawn into the system, but this outdoor air for ventilation must also be delivered to where the building occupants are. This section of this chapter focuses on the determination of the OA quantity entering the HVAC system, while a latter section of the chapter focuses on the determination of the OA quantity delivered to the occupants.

The techniques for determining the OA quantity entering an HVAC system include either a direct measurement of the volume of this air stream, or measurement of the total air volume in combination with a determination of the portion of this total supply air volume that is outdoor air. In this second approach, the percentage of the percentage of outdoor air can be determined by measuring some component which varies among the three air streams (outdoor air, return air, and the mixed or supply air), such as its enthalpy, temperature, carbon dioxide (CO_2), or tracer concentration. It should be noted also that the determination of the OA quantity reaching the building occupants can also be achieved by CO_2 measurements, tracer measurements, or by the combination of direct measurement of the volumes of air supplied to the location in conjunction with a determination of the percentage of outdoor air in that supply air.

Quantity of Outdoor Air Entering the HVAC System

In making the determination of the OA quantity that is being drawn into this HVAC system, consideration should also be given to the qualility of outdoor air. However, before either of these questions can be answered, there needs to be an understanding of the influence of the design, installation, operation, and maintenance of the systems intended to provide the outdoor air.

When evaluating the OA quantity entering the HVAC system, there needs to be an appreciation of the fact that the following three situations can exist.

1. All of the outdoor air entering the HVAC system may not necessarily be coming in through the outdoor air intake. In this situation, some may be coming in at other locations and passing through the mechanical room.
2. The OA quantity entering the HVAC equipment may not comprise all of the outdoor air entering the building, due to infiltration at other locations.
3. The OA quantity entering the HVAC equipment may not represent the OA quantity being delivered to the building occupants, due to distribution inefficiencies.

RELATIONSHIP BETWEEN OUTDOOR AIR QUANTITY AND THE HVAC SYSTEM

The relationship between the OA quantity both entering the HVAC system and being delivered to the building occupants is a function of all of the factors that have the potential to affect the HVAC system or systems. The major factors of the HVAC system can be broken down, for discussion purposes, into four categories: (1) its design, (2) its installation, (3) its operation and (4) its maintenance. This distinction can be arbitrary however. For instance, if a system is designed so that it is difficult for maintenance to be performed and, as a result, poor maintenance practices contribute to an IAQ problem, it is an open question as to which category this situation belongs to. With the exception of this type of crossover situation, the following discussion presents the role of the these factors in their relationship to outdoor air quantities.

Design Factors Affecting Outdoor Air Quantity

The underlying design of the HVAC system has a major impact on the OA quantity that can delivered to the building occupants. For instance, in the late 1970s many buildings were designed and built with an overriding emphasis on energy conservation and minimization of first costs. Many buildings constructed at this time therefore have HVAC systems that can provide only a constant, minimum OA quantity. In contrast, please note that there are also other buildings whose designers recognized the energy-saving potential of air economizers. The buildings with fixed minimum outdoor air control can either be without or with a return air fan. For systems *without* return fans, the outdoor air damper (see Figure 5.1) is interlocked to open only when the supply air fan operates. The actual OA quantity entering will be a function of the damper opening and the pressure difference between the mixed air plenum and the outdoors. For buildings where there is not a direct provision for building relief, or exhaust at the AHU (air handling unit), the OA quantity entering the building will ultimately only be equal to the quantity of air leaving the building due to exhausts (such as for the bathrooms) and exfiltration less the infiltration at other locations.

For buildings with fixed minimum outdoor air control with return fans, the situation becomes more complex because the outdoor air quantity, minimum or otherwise, also depends on the airflow difference between the supply and return fans, as well as the OA damper position, the other building exhausts, and amount of exfiltration and infiltration.

Another design approach for controlling the OA quantity entering the HVAC equipment is called "economizer control," where the OA quantity can increase as a function of the outdoor air temperature and thereby provide "free cooling." There are two types of economizer systems: temperature economizer and enthalpy economizer. With temperature-based economizers, the OA quantity to be introduced is based on the dry-bulb, or sensible, temperature of the OA stream. Typically, this latter system operates with an outdoor air control regime based on achieving a constant mixed air temperature. In this control regime, if the outdoor

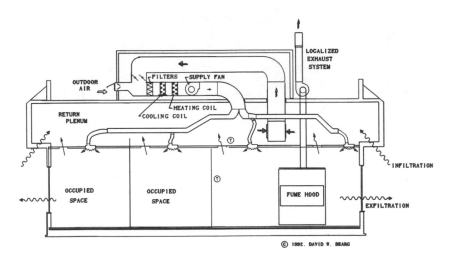

Figure 5.1   HVAC system with fixed minimum outdoor air, without return air fan.

air temperature is below a high-temperature limit, typically anywhere between 65°F and the temperature of the return air (RA) stream, the return, exhaust, and OA dampers (see Figure 5.2) modulate to maintain a mixed air temperature close to the desired delivery temperature for the building. There will, of course, be a temperature rise of a few degrees as the mixed air passes by the supply air fan, picks up heat from the fan motor, and becomes the supply air. When the outdoor air temperature exceeds the high-temperature limit setpoint, the OA damper reverts to its fixed minimum position and the return air damper goes to full open.

Example 5.1. For a HVAC system operating with a return air temperature of 75°F, an economizer high-temperature limit of 75°F, a minimum outdoor setting of 20%, and a mixed air goal of 58°F, the resulting relationship that can be expected to exist between the outdoor air temperature and the percentage of outdoor air is presented in Table 5.3.

As can be observed from the data in Table 5.3, economizer controls have the dual advantage of being able to help maintain good indoor air quality and minimize energy costs at the same time.

When evaluating buildings with economizer capability, the OA quantity entering the HVAC equipment can therefore vary during the day as the outdoor air conditions change. Therefore, to maximize the usefulness of the IAQ evaluation data, it is recommended that the controls of the system be overridden or reset to achieve the minimum outdoor air conditions even when they might not otherwise be expected to occur. One way of achieving the minimum outdoor air condition would be to reset the economizer high-temperature sensor/controller to a set point

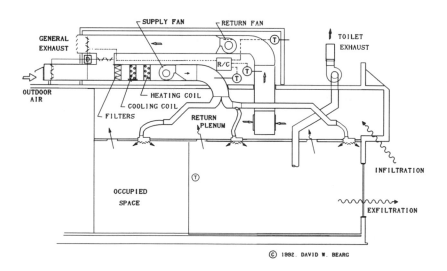

© 1992. DAVID W. BEARG

Figure 5.2   HVAC system with economizer control of outdoor air quantity.

Table 5.3.   Estimate of %OA as a Function of Outdoor Air Temperature

Outdoor Air Temperature °F	%OA
>75	20
70	100
65	100
60	100
55	85
50	68
45	57
40	49
35	43
30	38
25	34
20	31
15	28
10	26
5	24
0	23

below the outdoor air temperature. The results will then indicate a "worst case" condition for the building, as opposed to a condition that can merely be described as typical or even "those conditions that existed on the day of testing." This situation is a perfect example of the need to understand the operating conditions of the HVAC systems before IAQ measurements are taken.

The second type of economizer control, enthalpy based, attempts to further reduce energy costs for cooling by considering the amount of humidity in the outdoor and return air streams instead of just their temperatures. The inclusion of enthalpy, the internal energy of the air/moisture mixture, takes into consideration the latent as well as sensible heat of these air streams. Enthalpy, which is expressed as BTU per pound of dry air, increases both with increasing moisture content and increasing temperature. The result is that although the return air temperature may be higher than the outdoor air temperature, because the outdoor air is at a higher dew point, less energy would be consumed in cooling the return air than the outdoor air. Other control regimes which affect the OA quantity entering the HVAC equipment are warm-up controls and night cool-down controls. With warm-up controls, the OA dampers remain closed as the building is heated from its nighttime setback temperature up to its occupancy setpoint. Ventilation systems for spaces with intermittent or variable occupancy may have their OA quantity adjusted by the use of dampers or by stopping and starting the fan system to provide sufficient dilution to maintain contaminant concentrations within acceptable levels at all times. According to *ASHRAE Standard 62-1989*,[10] these system adjustments may lag or should lead occupancy depending on the source of contaminants and the variation in occupancy. When contaminants met the following conditions:

1. They are associated only with occupants or their activities.
2. They do not present a short-term health hazard.

3. They are dissipated during unoccupied periods to provide air equivalent to acceptable outdoor air.

then the supply of outdoor air may lag occupancy. Conversely, when contaminants are generated in the space of the conditioning system independent of occupants or their activities, the supply of outdoor air should lead occupancy so that acceptable conditions will exist at the start of occupancy. This standard provides charts that can be used to calculate permissible lag and lead times.

In addition to the various outdoor air control approaches, another aspect of the building and HVAC design is whether the HVAC system is based on a centralized mechanical room with an extensive system of ductwork for distributing the supply air or merely a decentralized system of unit ventilators distributed around the perimeter of the building. Depending on which situation is present, the approach of the investigation will differ by focusing on the mechanical room or a representative number of distributed units.

Installation Factors Affecting Outdoor Air Quantity

Beyond the intentions of the design, the details of the actual installation also affect the ability of the HVAC system to deliver adequate quantities of outdoor air to the building occupants. In one building, the HVAC system was based on a packaged system located on the roof. In this package system, all of the wiring had been done at the factory except for the wiring for the building relief fan, which was performed in the field. This on-site portion of the installation, however, was performed incorrectly and the building relief fans failed to operate. This installation problem led to inadequate quantities of outdoor air being drawn into the building because, without the operation of these fans, there was insufficient pressure drop at the OA dampers to draw in the intended air volume. As stated elsewhere, it is the combination of both the net open area at the dampers plus the pressure drop across this opening that determines the OA quantity entering the AHU. Other installation errors include the placement of VAV (variable air volume) mixing boxes upside down from their intended arrangement. Since the damper position was intended to work with gravity to keep it open, its being upside down led to the damper remaining closed when it wasn't supposed to be.

Operational Factors Affecting Outdoor Air Quantity

There are many operational parameters than can affect the OA quantity entering the HVAC equipment. For instance, although a given system may have been designed to be capable of regulating the outdoor air quantity with an economizer cycle, it may not be operated to take advantage of that fact.

Example 5.2. In one building in which an IAQ evaluation was being performed, the information provided by the engineering staff of a large research and manufacturing campus at a preinspection meeting indicated that the building in question had an economizer cycle and, based on the current outdoor air temperature, would have it's OA dampers wide open on the day of our inspection. Upon inspection, however, the OA dampers were observed to be open only to their minimum position. It turned out that, unbeknownst to his bosses, the individual responsible for that particular building had overridden the design intent of the controls of the HVAC system. The reasons for his decision relate to issues of design and maintenance; the design placed spray nozzles for humidification in the mixing chamber, upstream of the filters, and were therefore being subjected to frequent maintenance due to problems of freeze damage to the spray nozzles. This example points out the interrelationship between design, maintenance, and operation of the HVAC in achieving good IAQ.

In addition to this example, where the outdoor air quantity was reduced to prevent freeze-ups of humidification spray nozzles, concern about the freeze-up of coils in the AHUs is a similar problem. There is a sensor/controller called a "freeze stat" which measures the temperature of one point in the air stream, typically between the filter bank and the first set of coils. If the measured temperature at this location fails below the setpoint of this controller, typically 45°F, then either the OA dampers close or the fan shuts down "on freeze stat." Depending on the representativeness of the sampling location for this sensor, as well as the calibration of this sensor, the system may shut down when the situation does not technically justify it.

Example 5.3. Another building where the design affected the operation of the HVAC system also had the inherent capability of an economizer cycle; however, it wasn't being utilized. In this building, the determining factor for the position of the OA dampers was not the outdoor air temperature, but in fact was the need to keep a certain minimum load on the chillers. Therefore, numerous opportunities to utilize "free cooling" and increase the ventilation rate in the building were not taken advantage of because the OA dampers were being manually kept at their minimum position to maintain the load on the chiller.

Example 5.4. In another building, complaints of inadequate ventilation were high on Mondays and Tuesdays, but dropped off dramatically for the rest of the week. The investigation subsequently determined that the time clock used to provide reduced ventilation during the weekend days was not set up correctly, so that Monday and Tuesday received the reduced condition intended for Saturdays and Sundays. This situation is an example of the need to have adequately trained personnel operating HVAC systems.

Another situation to be aware of involves the use of "duty cycles," where fans are periodically shut off to conserve energy. If this practice is occurring when the building is occupied, it represents a violation of building codes.

Another concern is that many building HVAC systems operate only until 5:00 or 6:00 p.m. in the evening, despite the fact that there is significant occupancy in the building until 9:00 or even 11:00 p.m. Frequently, in these situations, there may be a mechanism for a tenant to have the ventilation system remain on longer at their request and expense. However, there may be a reluctance for them to incur this expense as often as they should. *Therefore, the operational schedule for the outdoor air quantity in relation to the occupancy patterns should be determined as part of an IAQ evaluation.*

After determining how the HVAC system is scheduled or programmed to operate, it should be verified that it is actually operating as intended. The system may not have any sensors or controls to monitor OA quantity; most do not. In most buildings, the OA quantity is usually only estimated, or determined indirectly by temperature ratios, or fan speed and static pressure.

The OA quantity entering the HVAC system may vary considerably over time. It is therefore important to assess this quantity under the minimum outdoor air and ventilation flow conditions. It is under these conditions that problems are likely to occur or intensify, unless an important source of air contaminants is located outside.

Maintenance Factors Affecting Outdoor Air Quantity

The maintenance issues that have the potential for affecting the OA quantity entering the dampers are more limited than those categorized as due to design, installation, or operation. Maintenance activities are more likely to affect the air contaminants side of the equation rather than the ventilation quantity. This subject is discussed in Chapter 10. There is, however, one maintenance activity — that of keeping the OA intake unclogged — which can directly affect the OA quantity entering the HVAC system. Therefore, the presence of obstructions in grilles or screens of air intakes should be visually assessed. Dirty coils will also reduce the overall quantity of supply air; and with it, potentially the OA quantity as well.

DETERMINATION OF THE QUANTITY OF OUTDOOR AIR ENTERING VIA THE OUTDOOR AIR DAMPERS

The OA quantity entering the OA dampers is a function of the combined influence of two parameters: (1) the net open area of the dampers, and (2) the pressure differential across these dampers. Therefore, the evaluation/inspection should note both the position of the OA damper and the pressure differential that exists across this air intake opening at the time of testing. Note, the pressure difference will vary as a function of the position of both the OA and RA dampers.

The actual determination of the OA quantity entering an AHU can be achieved by either a direct measurement of the air flow, or indirectly using the measurement of either the return air or mixed air volume plus the measurement of some attribute of the three air streams involved, the OA, RA (return air), and MA (mixed air).

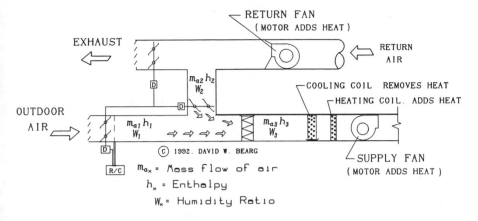

Figure 5.3   Adiabatic mixing of two air streams.

Direct measurement of the outdoor air quantity, in cfm, can be accomplished by determining the average flow velocity in a duct or across an opening, and then multiplying this value by the cross-sectional area of the duct or the net open area of the opening. This is much easier said than done, due to the very short distance that typically exists between the outdoors and the mixing chamber of the AHU. Therefore, the determination of the outdoor air quantity is usually based on direct measurement of either the supply air (SA) volume or the RA volume and an apportionment, according to some property, of the three air streams, such as temperature, enthalpy, CO_2, or tracer concentration.

Temperature-Based Determination of Percentage Outdoor Air

The mixing of the RA stream with the OA stream to create the mixed air, and ultimately the supply air, stream can be depicted as shown in Figure 5.3.

The three air streams can be characterized by several parameters, including volumetric flow rate, temperature, humidity ratio, and enthalpy. If the RA and the OA streams combine without gaining or losing heat (i.e., adiabatically), then the laws of conservation of energy will permit the calculation of the percentage of outdoor air. The equations that describe this relationship can be summarized as follows.

$$\frac{h_2 - h_3}{h_3 - h_1} = \frac{W_2 - W_3}{W_3 - W_1} = \frac{m_{a1}}{m_{a2}} \tag{5-1}$$

where
h_i = enthalpy of air stream i, Btu/pound
W_i = humidity ratio of air stream i, lb of water/lb dry air
m_{ai} = mass flow of dry air stream i per unit time

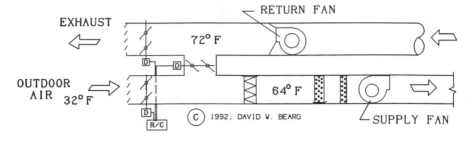

$$\frac{RA-SA}{RA-OA} = \frac{72-64}{72-32} \times 100 = \frac{8}{40} \times 100 = 20\% \ O.A.$$

Figure 5.4 Calculation of % outdoor air from temperature measurements.

Fortunately, however, the need to determine the actual enthalpy values is not necessary to perform this apportionment since the same relationship will exist between the sensible temperatures as it does with the enthalpies. The correspondence between the enthalpies and dry-bulb temperatures can be observed on a psychrometric chart. The equation to calculate the ratio of the OA flow to the RA flow can therefore be approximated as:

$$\frac{t_2 - t_3}{t_3 - t_1} = \frac{m_{a1}}{m_{a2}} \qquad (5\text{-}2)$$

where t_j = dry-bulb temperature of airstream j, Fahrenheit.

For ease of computation of the percentage of outdoor air (%OA), the following equation can also be used.

$$\frac{t_{RA} - t_{MA}}{t_{RA} - t_{OA}} \times 100 = \% \ \text{Outdoor air} \qquad (5\text{-}3)$$

An example of the application is provided in Figure 5.4, where the outdoor air temperature is 32°F, the return air temperature is 72°F, and the mixed air temperature is 64°F; these values are used to yield a value of 20% outdoor air.

While this procedure appears straightforward, it is very difficult to obtain reliable information from this technique in actual buildings. The first primary obstacle involves the difficulty in obtaining a correct value for the mixed air temperature. If there is a large temperature difference between the OA and RA streams, then they will have differing densities which will lead to poor mixing between them. Figure 5.5 displays the results of an actual 42-point measurement distribution of mixed air temperatures that were obtained downstream of the filter bank and just upstream of the cooling coils.

55.6	54.6	56.3	58.5	58.0	53.7	50.2
55.8	54.2	57.6	63.8	59.5	59.0	49.4
58.0	62.4	63.3	66.4	67.8	61.2	56.3
62.3	65.8	66.4	68.7	67.6	64.7	58.8
63.5	65.0	66.0	67.4	67.4	63.9	61.4
63.6	64.8	65.4	66.8	65.7	62.4	60.6

Outdoor air temperature   = 30.3°F
Return air temperature   = 72.5°F
Average mixed air temp.   = 61.4°F
Mixed air S.D.   = 4.9°F
% OA: = 26.2% (Based on Average MAT)
                       14.5% (Avg MAT + 1 S.D.)
                       38.0% (Avg MAT − 1 S.D.)
(S.D. = standard deviation).

Figure 5.5   Distribution of Mixed Air Temperature Values, °F

The face of the cooling coils measured 9 ft across by 8 ft 9 in. high, for a total face area of 78.75 ft^2. Each of the 42 measurements therefore needs to be representative of 270 in^2, an area equivalent to an opening 18 × 15 in. Therefore, not only is there some uncertainty as to the representativeness of each of the individual measurements; but there is also significant uncertainty as to the correct average value to use, since the standard deviation of 4.9°F corresponds to a range of possible averages from 56.5 to 66.3°F. The resulting range of %OA, which averages 26.2%, can actually be from 14.5 to 38%.

In this situation, if the measurement of the mixed air temperature was based on the just one measurement, then the potential for error is even greater. Of the 42 measurements, the highest value was 68.7°F, 1.5 standard deviations from the mean. Calculating the %OA from this one measurement yields a value of 9% OA. At the other extreme, the lowest of the 42 measurements was 49.4°F, 2.45 S.D. from the mean. Calculating the %OA from this one measurement yields a value of 55% OA.

Another interesting piece of information that can be gleaned from this data is that although the colder air is denser and heavier than the warmer air, the lowest temperatures are not observed along the bottom of the coil face. A swirling action has in fact brought some of the coldest air up near the top. The significance of this

Table 5.4.   Estimate of Percentage Outdoor Air, Case 2

Return Air Temperature   = 72.5°F
Outdoor Air Temperature = 74.0°F

Mixed Air Temperature °F	% Outdoor Air (%OA)
73.5	66.67
73.4	60.00
73.3	53.33
73.2	46.67
73.1	40.00
73.0	33.33
72.9	26.67
72.8	20.00
72.7	13.33
72.6	6.67

is that the placement of single-point thermal sensors should be preceded by measurements of the mixing characteristics to ensure that they are indeed representative of what they are intending to measure.

However, the mixing box geometries of all HVAC systems cannot be expected to yield as poor mixing conditions as this particular example. The purpose of this information is to raise the issue of how variable this data can be and that a sufficient number of measurements need to be taken to achieve a level of confidence in the results.

The other potential situation when attempting to perform temperature-based determination of the %OA is that there is not a large temperature difference between the OA and RA streams; then, although there will be better mixing of the two air streams, the small differences in temperature result in large computational sensitivity to small errors in measurement. For instance, when the outdoor air temperature exceeds the return air temperature, attempting to quantity the %OA value demands precision of measurement to a tenth of a degree. The sensitivity of this measurement on the calculation of the %OA value is presented in Table 5.4.

As can be observed in Table 5.4, a 0.2°F difference in measurement between 72.7 and 72.9°F corresponds to a difference in the computed value of %OA from 13.3 to 26.7% — not a reliable or practical assessment technique.

Another limitation of this temperature-based approach is that there are many AHUs where it is physically impossible to physically gain access to a location where the OA and RA streams have mixed and yet have not had heat added or taken away. Once the air stream has passed through an active coil, either heating or cooling, this technique cannot be applied. Even the supply air fan adds heat to the air stream. For these reasons, the use of CO_2 or sulfur hexafluoride (SF_6) tracer measurements are more powerful tools for the determination of the %OA.

Carbon Dioxide-Based Determination of Percentage Outdoor Air

The CO_2-based technique for determining the %OA entering the AHU is similar to the temperature based approach in that it relies on the conservation of

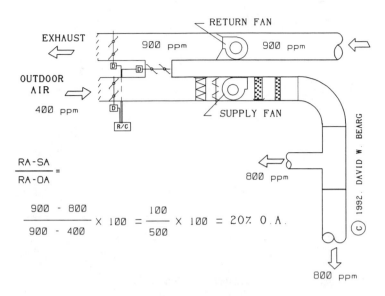

Figure 5.6   Carbon dioxide calculation of % outdoor air.

mass and energy. CO_2 is neither created nor destroyed in the mixing of the OA and RA streams to form the SA stream. In this approach, the measurement of the CO_2 concentration can be obtained at any SA register in the building. It is prudent in this situation, however, to collect several measurements at differing locations, to confirm that the CO_2 has become uniformly mixed and distributed within the air stream. Equation 5-4 can be used for the calculation, and an example is presented in Figure 5.6.

$$\frac{CO_{2-RA} - CO_{2-MA}}{CO_{2-RA} - CO_{2-OA}} \times 100 = \% \text{ OA} \qquad (5\text{-}4)$$

Example calculation from Figure 5.6:

$$\frac{900 - 800}{900 - 400} \times 100 = \frac{100}{500} \times 100 = 20\% \text{ OA}$$

This example is straightforward enough, until the additional information is given that the outdoor air concentration of 400 ppm was obtained at the OA intake. Another measurement obtained at the upwind face of the roof was 350 ppm. The difference in outdoor air concentration of CO_2 is due reentrainment of the exhaust from the building, typically where the exhaust plume is trapped in the aerodynamic wake of the building and a portion of these gases reenter the building via the air intake. The aerodynamic wake of a building refers to the areas of recirculation around the building, especially on the roof, which are created by

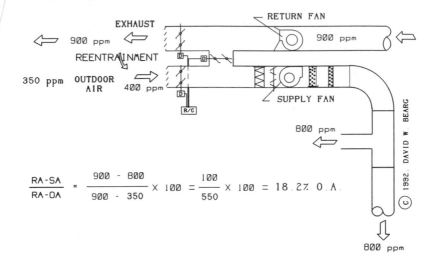

Figure 5.7   Carbon dioxide recalculation of % outdoor air.

$$\frac{RA-SA}{RA-OA} = \frac{900-800}{900-350} \times 100 = \frac{100}{550} \times 100 = 18.2\% \ O.A.$$

$$\begin{bmatrix} 900 \ ppm \\ -350 \ ppm \end{bmatrix}$$
$$\triangle = 550 \ ppm$$

EXHAUST   900 ppm

CLEAN
OUTDOOR AIR
$$\begin{bmatrix} 350 \ ppm \\ -350 \ ppm \end{bmatrix}$$
$$\triangle = 0 \ ppm$$

9.1%   OUTDOOR AIR
WITH REENTRAINED
EXHAUST AIR   400 ppm

$$\begin{bmatrix} 400 \ ppm \\ -350 \ ppm \end{bmatrix}$$
$$\triangle = 50 \ ppm$$

90.9%

INCREMENT
ABOVE AMBIENT

$$(.909) X \ 0 \ ppm + (.091) X \ 550 \ ppm = 50 \ ppm$$

Figure 5.8   Carbon dioxide determination of reentrainment.

the interaction of the wind and the building. Reentrainment can also occur when there is the direct flow of an exhaust plume to the capture zone of the air intake. The recalculation of the %OA including this condition is presented in Figure 5.7.

$$\frac{900-800}{900-350} \times 100 = \frac{100}{550} \times 100 = 18.2\% \ OA$$

Quantification of the amount of reentrainment can also be determined from this data. This computation is simplified if the background concentration of CO_2 is subtracted from each of the values. The resulting values are presented in Figure

5.8. Having eliminated the background concentration from the calculation, the 50 ppm CO_2 now showing up at the OA intake must be entirely due to the reentrainment. Thus, dividing the 50-ppm value by the 550 ppm (above ambient) of the return air, and therefore of the exhaust plume, results in the determination that 9% of the outdoor air is really exhaust air reentering the building.

Sulfur Hexafluoride-Based Determination of Percentage Outdoor Air

The calculations used in the SF_6-based determination of the %OA in the supply air are almost identical to those used with CO_2. This is because, in both situations, these computations are comparing concentrations of one component of the air stream. The advantage of the SF_6 approach over that of the CO_2 approach is that greater precision that can typically be achieved in the measurements. For instance, the small amount of outdoor air coming in via a unit ventilator would be impossible to quantify with a CO_2 detector that only outputs concentrations to the nearest 50 or 25 ppm. Also, the computation for the amount of reentrainment occurring is simpler because the background concentration does not need to be subtracted out first. The equation to use for the SF_6 approach is therefore:

$$\frac{SF_{6-RA} - SF_{6-MA}}{SF_{6-RA} - SF_{6-OA}} \times 100 = \% \, OA \qquad (5-5)$$

The use of this Equation 5-5 is presented in Figure 5.9. In this example, the calculation of the %OA distinguishes between the *actual* quantity of "fresh air"

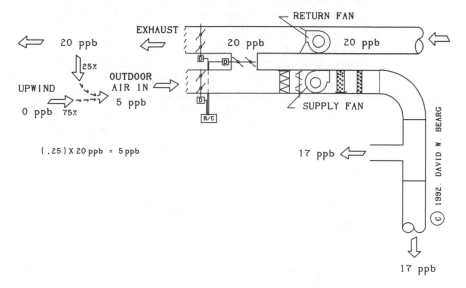

Figure 5.9 SF_6 determination of % outdoor air and reentrainment.

entering the building (and includes the impact of reentrainment) and the *apparent* quantity of "fresh air" entering the building (which ignores the impact of reentrainment by assuming that the outdoor air is tracer-free).

Actual %OA quantity is given by:

$$\frac{20-17}{20-0} \times 100 = \frac{3}{20} \times 100 = 15\% \text{ OA}$$

Apparent %OA quantity is given by:

$$\frac{20-17}{20-5} \times 100 = \frac{3}{15} \times 100 = 20\% \text{ OA}$$

This specific computation therefore reflects the dynamics of the position of the OA dampers and the pressure differential across them. Both of these computations are important. The "actual" value will vary as a function of the wind speed and direction and therefore may relate to IAQ problems that exist only during certain wind regimes. The "apparent" value, however, should remain relatively constant over time. Any variation in the apparent value can be caused by changes in the infiltration and exfiltration across the building envelope due to differing tempera-ture differences and wind speeds. In modern, well-sealed buildings, however, these variations tend to be minor.

If there were no reentrainment occurring, then the outdoor air concentration of tracer would effectively be zero, and the actual and apparent outdoor air quantities would be the same. However, since the outdoor air concentration was measured to be 5 ppb, the fraction of reentrainment can be calculated by dividing this concentration by the exhaust concentration of 20 ppb. In this example, therefore, the amount of reentrainment occurring is $1/4$, or 25%. The significance of this amount of the exhaust plume reentering the building is that 25% of the air at the OA intake is air that has already been through the building.

In this example, then, since 25% of the apparent outdoor air is really just recirculated air, the remainder, (1.0 to 0.25), or 0.75 of the apparent %OA (which in this case is 20%) is equal to the actual value of 15% OA. Reentrainment is an important issue because it can make the difference between a sufficient and an insufficient OA quantity entering an AHU.

OTHER PATHWAYS OF OUTDOOR AIR INTO THE HVAC EQUIPMENT

The OA quantity entering the HVAC system may not comprise all of the outdoor air entering the building and all of the outdoor air entering the HVAC system may not necessarily be coming through the OA intake. The importance of these two facts can be illustrated by a building evaluation where the occupants in one location in a building were complaining that they were getting exposed to diesel exhaust fumes from vehicles idling at the loading dock, one floor below their location. In response to these complaints, the air intake for the HVAC system serving this and other locations was relocated, and yet the problem persisted. The geometry of the situation is displayed in Figure 8.12 in Chapter 8. This situation is presented in Chapter 8 as part of the discussion on assessing the pathways of air movement into buildings. Note in particular the fact that the mechanical room is resting on hollow precast concrete slabs which are located directly above the loading dock. Aided by the use of an air current tube, the visual inspection determined both that there were numerous penetrations between the loading dock and the mechanical room, and that the air was being drawn into the mechanical room. The next step in the evaluation was the release of a small quantity of tracer, SF_6, at one of the penetrations identified, followed by sampling and analysis for tracer in and around the office space where the complaints had come from.

As was the case in this example, mechanical rooms are typically under negative pressure with respect to the outdoors because the ductwork on the suction side of the fan is not air-tight. These leakage sites are due to both the presence of unsealed metal-to-metal joints and to gaps created by penetrations for controls. One of the worst examples of degraded IAQ due to the unintentional introduction of air contaminants via a mechanical room involved the presence of large holes in the flexible fabric seals connecting the ductwork to the fan housing. There were other contributing factors in this building: the original design called for penthouse air intake fans to pressurize the air intake shaft to transfer this air to the AHUs in the basement mechanical room. Unfortunately, however, it was decided not to operate the penthouse fans due to noise and vibration concerns. The fans in the basement AHUs were then sped up to overcome the increased pressure drop in the OA intake shaft. Tracer testing, as discussed in Chapter 8, documented the extent of the unintentional air intake from the loading dock.

In summary then, the inspection and testing effort of an IAQ evaluation needs to consider the outdoor air drawn into the HVAC equipment from the mechanical room itself, in addition to the outdoor air drawn in the AHUs at the OA dampers. The controlling determinants in this situation are the combination of two factors: (1) the construction of most AHUs does not achieve an air-tight condition, and (2) the fact that this air handling equipment, especially at the supply air fan, is being operated at a negative pressure (typically several inches of water column static pressure) with respect to the mechanical room. The magnitude of this unintentional leakage, as with the amount of air passing through the OA dampers, is a function of both the net open area and the pressure drop across that opening.

The significance, in terms of resulting IAQ, of the unintentional leakage of air from the mechanical room into the air handling equipment will depend heavily on the actual location of the mechanical room in the building. The following discussion therefore considers situations where the mechanical room is located in the rooftop penthouse, the basement or lower level, or at location in the middle of the building.

It should also be mentioned that from the perspective of minimizing reentrainment of air contaminants exhausted from a building, the optimum location for air intakes is usually at a level equal to one third of the height of the building.

Infiltration and Natural Ventilation

Infiltration and natural ventilation are really one and the same phenomenon; the distinctions for discussion purposes here are that infiltration is unintentional and may be influenced by pressure differences across the building envelope, while natural ventilation is intentional and is designed to supplement or eliminate the need for mechanical ventilation or to be available in an emergency if the mechanical ventilation system is not working. One fact that needs to be remembered with both of these situations is that, compared with mechanical ventilation systems, the outdoor air entering the building will be neither thermally conditioned nor filtered.

The driving force for the movement of air in these, and all situations, is the presence of a pressure differential between indoors and outdoors. In a heated building, the leakage of the warmer indoor air, being more buoyant than outdoor air, through penetrations at the upper portion of the structure create a partial vacuum or negative pressure in the building to draw in air to replace that which was lost to the outdoors. This situation, commonly known at the stack effect, needs to be included in the overall evaluation of the building and its IAQ, and is discussed in more detail in Chapter 8.

REFERENCES

1. Chan-Yeung, M. and S. Lam. 1986. "Occupational asthma." *Am. Rev. Respir. Dis.,* 133:686–703.
2. Burge, P. S., G. Edge, M. O'Brien, M. G. Harries, R. Hawkins, and J. Pepys. 1980. "Occupational asthma in a research center breeding locusts." *Clin. Allergy,* 10:355–363.
3. Figley, K. D. 1940. Mayfly (Ephemerida) hypersensitivity. *J. Allergy,* 11:376–387.
4. Gibbons, H. L., J. R. Dille, and R. G. Cowley. 1965. "Inhalant allergy to screwworm fly." *Arch. Environ. Health,* 10:424–30.
5. Bernton, H. S., T. F. McMahon, and H. Brown. 1972. "Cockroach asthma." *Br. J. Dis. Chest,* 66:61–66.
6. Harfi, H. A. 1980. Immediate hypersensitivity to cricket. *Ann. Allergy,* 44:162–163.
7. Stevenson, D. D. and K. P. Mathews. 1967. "Occupational asthma following inhalation of moth particles." *J. Allergy,* 39:274–83.
8. The Commonwealth of Massachusetts, Special Legislative Commission on Indoor Air Pollution. *Indoor Air Pollution in Massachusetts, Interim Report.* June 1988. 93 pp.
9. Course brochure of the EOHSI/CET, UMDNJ/DCHE, October 1991.
10. American Society of Heating, Refrigerating, and Air-Conditioning Engineers. 1989. *Standard 62-1989: Ventilation for Acceptable Indoor Air Quality.* Atlanta, GA.
11. Carrier Corporation. *The ABC's of Air Conditioning.* 1975. (Syracuse, New York: Carrier Corporation). 24 pp.

CHAPTER 6

Quantity of Outdoor Air Delivered to Occupants

OVERVIEW

This chapter focuses on perhaps the most important question to be answered as part of a ventilation-based indoor air quality (IAQ) evaluation: "How much outdoor air is actually being delivered to the building occupants?" As discussed in Chapter 4, there exists the minimum outdoor air requirements of *ASHRAE Standard 62-1989, Ventilation for Acceptable Indoor Air Quality*, as an evaluation criterion. For offices, this consensus standard lists a minimum of 20 cubic feet per minute (cfm) per person of outdoor air (OA) that should be delivered to building occupants.

Although there apparently is still a widespread belief that the quantity of air delivered to the building occupants can be determined by merely measuring the OA quantity entering the HVAC system at the OA dampers and dividing it by the maximum number of people in the building, my testing experience indicates otherwise. Therefore, while it is important to assess the OA quantity entering the HVAC equipment, in order to compare current performance with rated capacities, it is equally important to actually quantify the volume of outdoor air in fact reaching the people. This chapter presents and discusses the three basic techniques for determining the OA quantity that is actually delivered to the occupied areas in a building. These techniques are based on the measurement of the buildup of carbon dioxide (CO_2) concentrations, the decay of sulfur hexafluoride (SF_6) concentrations, and the determination of the supply air (SA) quantity in conjunction with a determination of the percentage of outdoor air (%OA) in this SA quantity.

Table 6.1   Occupancy Loading and Outdoor Air Quantities

Location Floor	Occupancy Loading		Outdoor Air		cfm/Person
	People	%	cfm	%	
3	274	34	4,900	22	18
2	240	30	4,350	19	18
1	268	33	1,650	7	6
B	28	3	11,600	52	414
Totals	810	100	22,500	100	28

Importance of Distribution System

There are several reasons why the OA quantity entering the HVAC system does not necessarily reflect the quantity that actually gets delivered to the building occupants. The reasons why these two quantities can differ can be summarized as follows.

1. A portion of the supply air can short-circuit directly into the return plenum, due to leakage from its ductwork, thereby preventing this air from reaching the people.
2. Short-circuiting of the supply air can also occur across the ceiling when the supply and return registers are placed too close together. This situation is frequently created or aggravated when a previously open area is divided into cubicles separated by partitions.
3. An additional distribution problem is caused when the delivery of the supply air goes to locations where there are no people, such as corridors and lobbies or exhaust systems.

There are numerous examples of each of these reasons. The situations that cause these conditions are discussed in this chapter.

In one building, most of the building's fresh air went to the basement where most of the people weren't. The specifics of this building are summarized in Table 6.1. One reason that the majority of the air was going to the basement was that there was 12,000 cfm of process exhaust leaving this location.

Of the three basic approaches that can be used, either singly, or in combination, to determine the OA quantity that is actually delivered to the building occupants, each has its own advantages and disadvantages. The measurement of the build-up of CO_2 concentrations indoors over outdoor levels depends on the presence of people in the building; the technique relying on the measurement of the rate at which a nontoxic tracer gas is diluted and removed from the space requires expensive equipment; and the combination of the measurement of the delivered supply air quantities and a determination of the percentage of outdoor air (%OA) in that supply air can be very tedious to perform if there are a lot of supply diffusers present. More detailed descriptions of these techniques follow.

CARBON DIOXIDE MEASUREMENTS

The measurement of CO_2 concentrations as a technique for determining the OA quantity actually delivered to the occupants of a building is discussed first because, if the building is occupied, this approach provides the greatest amount of information for the least effort.

The basis for this approach relies on the dynamic relationship between a buildup of CO_2, a component of the indoor air which is proportional to the number of people in the building, and the ability of the ventilation system to dilute and remove this component. The behavior of this component of the indoor environment will therefore be representative of the behavior, in terms of accumulation or clearance rates of other components of the indoor air in general and especially for those that originate from the presence of people. In order to use CO_2 measurements to determine the OA quantity being delivered to the occupants, the number of people present should therefore remain constant. If this condition is met, then the ventilation rate can be calculated from the rate of build-up of indoor CO_2 concentrations over those of the outdoors. This determination is most simple if the indoor concentration builds up to equilibrium conditions, which neither increase or decrease for some interval of time. If these conditions are achieved, then this peak equilibrium value can be directly converted to the effective ventilation rate, which is expressed in terms of cfm of outdoor air per person. This result is then directly comparable with the 20 cfm per person outdoor air requirement listed in *ASHRAE Standard 62-1989* for office spaces. If the number of people present is also known, then a value for the amount of outdoor air in cfm can be computed as well.

Carbon Dioxide Generation Rate

The basic assumption of this assessment technique is that on average, each of the building occupants will generate CO_2 at a known and consistent rate. According to Appendix D of *ASHRAE Standard 62-1989,* this generation is equal to 0.3 liters per minute[1] (0.0106 cfm of CO_2). This is the generation rate for an activity level of 1.2 Met units, which corresponds to the level of exertion typical office activities. Another published value for this generation by Persily and Dols,[2] however, report a value of 5.3×10^{-6} meters per second per person (5.3×10^{-6} m³/s/person). This generation rate is however equal to 0.318 liters per minute (0.318 l/min), or 0.0112 cfm, which is almost 6% greater. A third "representative value of CO_2 production by a sedentary individual who eats a normal diet is 0.011 cfm," according to *ASHRAE Fundamentals.*[3] These, however, are average values and will change as a function of the diet and activity of the individual. For instance, according to Balvanz et. al.,[4] CO_2 generation rates of 0.5 l/min (0.0177 cfm) are reported. The difference between these generation rates and those for sedentary individuals in commercial buildings can be explained by the pressurization of the cabin and the fact that the people are somewhat more excited.

Similarly, in a given building, if the level of activity were more strenuous than that of typical office work, then the metabolic rate would go up with a corresponding increase in the CO_2 generation rate. For the same ventilation rate then, this increased level of activity would therefore result in an increased build-up of CO_2 concentrations.

When people are present in the building long enough that a steady-state condition has been achieved, then the quantity of CO_2 leaving the space due to the ventilation is equal to the quantity of CO_2 being generated. If this equilibrium condition has not yet been achieved, then the CO_2 concentration is still building up and the quantity leaving is smaller than the quantity being generated.

The basic equation for computing this relationship is derived from the continuity equation to yield the following relationship:

$$C_{indoor}(t) = \frac{F \times 10^6}{V_{eff} I}\left(1 - e^{-It}\right) + C_{outdoor} \tag{6-1}$$

where:

$C_{indoor}(t)$	=	CO_2 concentration indoors at time t, ppm (parts per million)
F	=	generation rate of CO_2, ft^3/hour/person
10^6	=	conversion factor from vol/vol to ppm
V_{eff}	=	effective volume, ft^3
I	=	ventilation rate, air changes per hour (ACH)
t	=	time, hours
$C_{outdoor}$	=	CO_2 concentration outdoors, ppm

One of the assumptions in this equation is that the generation rate, the ventilation rate, and the outdoor concentration remain constant over the evaluation interval.

One aspect of this equation is that, as the value of t increases, the (e^{-It}) term goes to zero so the $(1 - e^{-It})$ term becomes equal to unity and can be dropped from the equation. In a practical sense, when this term goes to unity, then equilibrium, (or steady-state) conditions have been achieved. As can be observed from this equation, the time required to achieve steady-state conditions is a function of the ventilation rate, I.

A useful calculation is to estimate how long it will take for these equilibrium conditions to be achieved. Recognizing that the assumed CO_2 generation of 0.3 liters per minute per person (0.3 l/min/person) has a potential error of 5%; a reasonable goal for the equilibrium calculation is the time required until the build up in the CO_2 concentration is within 5% of its peak value. Since $e^{-3} = 0.04975$, then $(1 - e^{-3})$ will equal 0.95. Therefore, the product of the ventilation rate times the time t, in hours, should be equal to 3 if equilibrium conditions are to be achieved. Therefore, at 2 ACH of outdoor air, near steady-state conditions will be achieved in 1.5 hours; that is how long the ventilation rate and the number of people present should remain constant. Similarly, at 1 ACH, the time required is 3 hours; at 0.5 ACH, the time required is 6 hours; and at 0.25 ACH, the time

Table 6.2.   Carbon Dioxide Generation Rate

Activity	Liters/minute	ft³/minute	met
Resting	0.20	0.0071	0.8
Sitting	0.25	0.0088	1.0
Light work	0.38	0.0135	1.2
Manual work	0.50	0.0177	1.6

required is 12 hours. At these lower ventilation rates, therefore, waiting for equilibrium conditions to be achieved is not practical. In addition, assuming that a peak afternoon measurement of CO_2 concentration is an equilibrium value when it is in fact not, and then using this value to calculate the ventilation rate will lead to an inflated value for the ventilation rate.

Assuming for a moment, however, that steady-state conditions have been achieved, then Equation 6-1 can be rewritten in the simplified form as presented in Equation 6-2. The derivation of Equation 6-2 from Equation 6-1 involves the subtraction of the $CO_{2outdoor}$ value from both sides, the change from a time value of hours to that of minutes, leading to the conversion of the ventilation rate, in ACH, times the effective volume term to the ventilation rate expressed in cfm/person:

$$\frac{cfm}{person} = \frac{10,600}{\left(C_{indoor} - C_{outdoor}\right)_{ppm}} \tag{6-2}$$

Remember that the value of 10,600 was obtained by converting the CO_2 generation rate of 0.3 l/min/person to American units of 0.0106 ft³/min/person and multiplying this value by 1 million to permit the direct use of ppm values of the CO_2 concentrations instead of the volume/volume form. In the volume/volume format, 350 ppm needs to be expressed as the number 0.00035. Another estimate of the average CO_2 generation rate is 0.011 ft³/min/person. Using this value instead leads to the following relationship:

$$\frac{cfm}{person} = \frac{11,000}{\left(C_{indoor} - C_{outdoor}\right)_{ppm}} \tag{6-2a}$$

The actual CO_2 generation for a given individual will depend on his diet and level of activity. Table 6.2 presents another set of values of the expected variation in CO_2 per person as a function of their activity level.

Getting back to the situation where equilibrium conditions have not yet been achieved before there is a significant change in the number of people present in the space or building, the ventilation rate can still be determined from measurements of the CO_2 and Equation 6-1. This calculation, however, must be done as an iterative process because a ventilation rate term appears on both sides of the equal sign.

Assumption of Steady-State Conditions

If it is assumed that a measured CO_2 concentration represents an equilibrium value, when in fact it is not, then the results will overestimate the amount of ventilation being provided. After people arrive in a building, the CO_2 concentration will build up through the day until the people begin to leave and the source strength for the CO_2 decreases. The peak value reached may, or may not, represent a steady-state condition. If the people stayed longer, higher concentrations might have been reached. The magnitude of the error in assuming that a peak value is a steady-state value increases as the ventilation rate decreases. Two example calculations are provided to show the extent of this potential error. In Table 6.3, the % error is calculated for the situation where the CO_2 concentration measured after a 3-hour occupancy is assumed to represent an equilibrium value; that is, it would not continue to increase if the people stayed longer. In this table, the ventilation rate value varies from 1.5 ACH of outdoor air down to 0.2 ACH of OA air. The assumptions used in this calculation are that the outdoor air has a CO_2 concentration of 350 ppm, there are 7 people per 1000 ft^2 of space, and the effective ceiling height is 10 ft.

Included in Table 6.3 are the ventilation rates expressed in the units of both ACH and cfm/person. This is followed by the value of the time constant, $(1 - e^{-It})$, for this ventilation rate at 3 hours. After this term is the expected CO_2 concentration value and then the calculated ventilation rate, in cfm/person, that would result from assuming that the measured CO_2 concentration was an equilib-

Table 6.3.   Assessment Error Inherent in Assuming CO_2 Reading Obtained After a 3-Hour Build-Up Represents Equilibrium Value
Case 1: High-Density Condition

Ventilation Rate		$(1-e^{-It})$ @ 3 hours	CO_2, ppm @ 3 hours	Calculated cfm/Person	% Error
ACH	cfm/Person				
1.5	35.7	0.989	643	36.1	1.2
1.4	33.3	0.985	663	33.9	1.6
1.3	31.0	0.980	685	31.6	2.1
1.2	28.6	0.973	711	29.4	2.9
1.1	26.2	0.963	740	27.2	3.9
1.0	23.8	0.950	773	25.1	5.3
0.9	21.4	0.933	811	23.0	7.3
0.8	19.0	0.909	856	21.0	10.0
0.7	16.7	0.878	908	19.0	14.0
0.6	14.3	0.835	969	17.1	19.9
0.5	11.9	0.777	1041	15.3	28.8
0.4	9.5	0.699	1127	13.6	43.2
0.3	7.1	0.593	1230	12.0	68.6
0.2	4.8	0.451	1354	10.6	121.8

Note: Occupant density = 7 people per 1000 ft^2; Effective ceiling height = 10 ft.; Volume/person = 1429 ft^3; $CO_{2outdoors}$ = 350 ppm.

rium value. The final column is the resulting error. For ventilation rates up to 1.1 ACH, or 26 cfm of outdoor air per person, this error is less than 4% and thus is considered acceptable.

A similar calculation is presented in Table 6.4; however, in this second table, the dilution volume per person is increased. Therefore, a comparison of Tables 6.3 and 6.4 indicates that with the larger dilution volume per person, a higher cfm per person value is computed for the same number of air changes per hour.

Example 6.1. In a densely packed building, with 1429 ft^3 per person (142.9 ft^2 in area by 10 ft in height), 1 ACH of ventilation is equivalent to 23.8 cfm/ person. By comparison, in a less densely packed building, with 2,200 ft^3 per person, the same 1 ACH of ventilation is equivalent to 36.7 cfm/person. As can also be observed from a comparison of these two tables, the CO_2 concentration values vary by both the ventilation rate and the occupant density, while the percent error varies only by the ventilation rate, expressed in terms of ACH of OA.

Advantages and Disadvantages

A point to remember is that the use of CO_2 measurements represents a powerful technique for determining the quantity of outdoor air delivered to the occupants of a building on a per person basis, provided that the investigator using this technique understands the limitations of this approach. Another useful technique for quantifying ventilation rates is the use of SF_6 tracer testing techniques. Again,

Table 6.4.   Assessment Error Inherent in Assuming CO_2 Reading Obtained after a 3-Hour Build-Up Represents Equilibrium Value

ACH	cfm/Person	$(1-e^{-lt})$ @ 3 hours	CO_2, ppm @ 3 hours	Calculated cfm/Person	% Error
1.5	55.0	0.989	540	55.7	1.2
1.4	51.3	0.985	553	52.1	1.6
1.3	47.7	0.980	568	48.7	2.1
1.2	44.0	0.973	584	45.3	2.9
1.1	40.3	0.963	603	41.9	3.9
1.0	36.7	0.950	625	38.6	5.3
0.9	33.0	0.933	649	35.4	7.3
0.8	29.3	0.909	678	32.3	10.0
0.7	25.7	0.878	712	29.3	14.0
0.6	22.0	0.835	752	26.4	19.9
0.5	18.3	0.777	799	23.6	28.8
0.4	14.7	0.699	855	21.0	43.2
0.3	11.0	0.593	921	18.5	68.6
0.2	7.3	0.451	1002	16.3	121.8

Note: Occupant Density = 5 per 1000 ft^2; Effective ceiling height = 11 ft.; Volume/person = 2200 ft^3; $CO_{2outdoors}$ = 350 ppm.

it is crucial that the investigator using this technique understands the requirements and limitations of this approach.

In summary, the advantages of the CO_2 measurement approach is the simplicity of the relationship between the number of people present and the build-up of CO_2 concentrations. When equilibrium conditions have been achieved, a simple equation permits the quantification of an estimate of the ventilation rate in the units of cfm of outdoor air per person, which is directly comparable with the minimum outdoor air recommendations of *ASHRAE Standard 62-1989*. Based on the uncertainty of the actual CO_2 generation rate occurring, this value will have a potential error of ±5%.

One shortcoming of this approach includes the difficulty in knowing whether equilibrium conditions have actually been achieved because if they have not and they are assumed to, then the results will overestimate the ventilation rate. Another limitation of this approach is the difficulty in knowing exactly how many people are present in the building at the time of testing. One reason for needing to know how many people are present during testing is that the test conditions may not represent the maximum potential occupancy of the building. For example, if testing is performed when only 75% of the potential population is present and the results indicate 21 cfm of outdoor air per person, the building would be considered as achieving the minimum requirements of *ASHRAE 62-1989*. However, by knowing that only 75% of the potential population is present, it is a simple computation to predict that at 100% occupancy, there would be less than 16 cfm of outdoor air per person being delivered.

CO_2 and Occupant Density

Since the amount of CO_2 being generated in a building is proportional to the number of people present, the fewer people present the less the CO_2 levels will build up. The following example illustrates this relationship.

Example 6.2. In a building with a floor area of 42,000 ft^2 and a total ceiling height of 11 ft (8.5 ft below the suspended ceiling plus 2.5 ft above the suspended ceiling), that is being ventilated at a rate equal to 0.1 cfm of outdoor air per square foot, the build-up rates for CO_2 are presented in Figure 6.1 for occupancies ranging from 2 to 7 people per 1000 ft^2 in area. For a building space with these dimensions, 42,000 ft^2 by 11 feet in height, the value for the gross volume would be 462,000 ft^3. 85% of this value, or 392,700 ft^3, would be the estimated effective volume. Applying the ventilation rate of 4,200 cfm of outdoor air to these values yields 0.55 and 0.64 ACH as the ventilation rates for these two categories of volumes. For this building space and the six different occupant densities, the resulting volume per person and the ventilation rate in cfm of outdoor air per person are presented in Table 6.5. In this table, the first column lists the population density in the number of people per 1000 ft^2 of office area. The second column lists the number of people that would be present in this example case of a 42,000 ft^2 area. The third column lists the value of the gross dilution volume, in cubic feet,

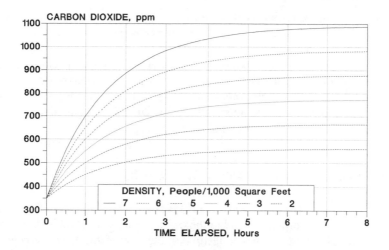

Figure 6.1   Buildup of CO_2 concentrations as a function of occupant densities .

Table 6.5. CO_2 Build-up Rate Data

Density	People	ft³/person	cfm/person	Peak CO_2
7	294	1571	14.3	1092
6	252	1833	16.7	986
5	210	2200	20.0	880
4	168	2750	25.0	774
3	126	3667	33.3	668
2	84	5500	50.0	562

available for each person. The fourth column lists the resulting ventilation rate in terms of cfm of outdoor air per person. The fifth column lists the peak CO_2 concentrations, in ppm, that would be measured if steady-state equilibrium conditions were achieved.

This example highlights the potential difficulties in using the CO_2 approach in buildings with low occupant densities, or that have been evacuated. *Measurements of CO_2, when used for ventilation rate deteminations, should always be accompanied by information on the number of people present and the duration of thier prior occupancy.*

Another important reason for knowing the population count is to be able to convert the (cfm of OA per person) value to a (cfm of OA) value so that this can be compared with the cfm of OA determined to be entering the HVAC system. This will permit a determination of how much of the OA entering the system actually gets delivered to the people. Another limitation of the CO_2 approach is that, typically, only one test condition can be evaluated in one day and the parameters of that test case need to be maintained consistently during the entire test interval. The major parameters involved are the number of people present plus the position of OA dampers and condition of the HVAC system. If these conditions vary during the testing interval, for instance, if the population count varies significantly or the position of the OA dampers changes, then the interpretation of the results are that much less definitive.

Owing to the uncertainties created by these reasons, it is useful and appropriate to employ a second approach, such at tracer testing, for determining the OA quantity delivered to building occupants. As discussed below, tracer testing allows greater control over the source strength and the time required for obtaining meaningful results is significantly reduced.

SULFUR HEXAFLUORIDE TRACER TESTING

The use of tracer gas measurements provides a very useful tool in the determination of the OA quantity delivered to the occupied areas of a building. The basic concept behind the use of tracer testing is the use of the measurement of a component of the air in a building in order to characterize some aspect of that building's performance or condition. Two basic references for this technique are the *ASTM Standard E741-83*,[5] which describes a standard tracer dilution method for infiltration measurement, and the discussion in *ASHRAE Fundamentals* (in Chapter 23 of the 1989 edition). The tracer most frequently introduced to perform this testing is SF_6. The characteristics of this gas that favor its selection include those listed in Table 6.6.

Although SF_6 has these desirable properties, it should be noted that toxic by-products may be formed when it is heated to decomposition (550°C).

It should also be noted that although SF_6 is much heavier than air, with a molecular weight of 146 compared with the average molecular weight of 29 for air, it does not settle out at the ppm or ppb (parts per billion) concentrations usually employed in IAQ studies. At these low concentrations, diffusion leads to uniform mixing with the rest of the air in the building.

Tracer gas measurements are based on a mass balance of the tracer gas within the building. This relationship is known as the continuity equation and, assuming that the outdoor concentration is zero, can be expressed as follows.

Table 6.6.   Desirable Characteristics of Tracer Gas for IAQ Testing

Measurable at very low concentrations
Inert, nonpolar, and not absorbed
Nontoxic, nonallergenic
Nonflammable and nonexplosive
Not a normal constituent of air
Measurable by portable equipment
Measurable by a technique that is free of interference
 by substances normally in air
Relatively inexpensive

$$V\left(\frac{dc}{dt}\right) = F - Qc \qquad (6\text{-}3)$$

where

c = tracer concentration, vol/vol
V = volume containing tracer, cu. ft.
F = tracer generation rate, cfm
Q = effective ventilation rate, cfm
t = time, minutes

In Equation 6-3, density differences between indoor and outdoor air are ignored; therefore the Q term, which can be referred to as the infiltration rate, includes both mechanical and natural ventilation in addition to envelope infiltration. The ratio of the effective ventilation rate, Q, to the volume, V, under evaluation has the units of volume/time, or air changes per time, and is called the air change rate, I.

Another requirement for tracer tests are that the tracer gas does not react chemically within the space and is not absorbed onto interior surfaces. There is also the assumption that the tracer gas concentration within the space is uniformly mixed and therefore can be expressed as one single value.

The determination of the OA quantity delivered to the occupied areas, herein referred to as the "effective ventilation rate" (EVR), can be achieved by three differing tracer gas methods: (1) the concentration-decay method, (2) the constant emission method, or (3) the constant concentration method. Of these three, the most frequently used for determination of the EVR is the concentration-decay, or tracer decay, method.

Tracer Decay Method

In this method, as explained in the *ASTM Standard E741-83*,[5] a small quantity of the tracer is released and thoroughly mixed into the space to be evaluated. This requirement of uniform mixing of the tracer is necessary for the results to be meaningful. After the release of the tracer has ceased, the F term in Equation 6-3 goes to zero and the now simplified relation can be written as it is in Equation 6-4.

$$C(t) = C_o \exp^{(-It)} \qquad (6\text{-}4)$$

where
$C(t)$ = tracer concentration at time t,
C_O = tracer concentration at time = 0
I   = air change rate, air changes per unit time
t   = time

Solving Equation 6-4 for the air change rate, I, yields Equation 6-5.

$$I = \left(\frac{1}{t}\right)\ln\left(\frac{C_o}{C}\right)$$

(6-5)

Equation 6-5 can also be used as it appears in Equation 6-6, where the tracer concentration data no longer has to begin with the measurement that was obtained at the time the tracer injection was terminated. In fact, it is preferable to wait 20 to 30 min after the injection has been terminated before calculating the air change rate.

$$I = \left(\frac{1}{\Delta t_{1\rightarrow 2}}\right)\ln\left(\frac{C_1}{C_2}\right)$$

(6-6)

For the results from Equation 6-6 to be in the units of ACH, then the time term, t, needs to be in the units of hours.

Plotting Tracer Decay Data

Before attempting to calculate a value for the air change rate from the tracer measurement data, I have found it most useful to first plot this data on semilog graph paper: the tracer measurements on the vertical (logarithmic) scale and time on the horizontal (linear) scale. This step is recommended because the shape of this line describes characteristics of the ventilation being provided. If this data yields a straight line, for instance, this verifies that the assumption of a well-mixed condition, inherent in this test procedure, has been achieved. Conversely, if this plot of the tracer data does not yield a straight line, then the characteristics of the ventilation system differ from the well-mixed case. Deviations from linearity can indicate either that the removal or air contaminants from the space is either occurring more efficiently, or less efficiently, than achieved by the well-mixed condition.

If the slope of the line deviates from linearity by having its concavity facing upward, then this result is indicative of the presence of a stagnant zone, such as that created by poor mixing of the supply air into the entire space. In this situation, compared with the well-mixed case, the removal of air contaminants is less efficient for the same volume of air being delivered.

If the concavity of the plotted line of the tracer decay data faces downward, indicating that the removal of air contaminants is more efficient than that occurring for the well-mixed case, then this is indicative of displacement or plug-flow ventilation. A more detailed discussion of these aspects of efficiency is presented in Chapter 7, Ventilation Characterization.

Another benefit provided by plotting the data, if multiple locations are being sampled, is that if the portion of the building being evaluated is not a single zone,

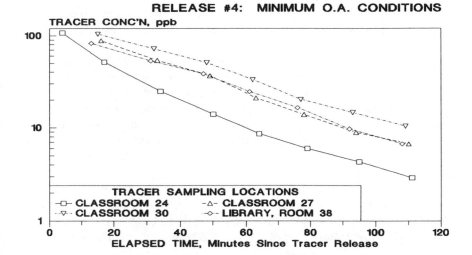

Figure 6.2   Tracer decay measurements for adjacent, differently ventilated locations.

but is in fact multiple interconnected zones, then the different sampling locations will yield different decay slopes that will tend to become uniform over time. When they do become uniform, then the slope of this line will represent the average ventilation rate for the entire volume, though quantification of the ventilation rates of individual portions of this zone will not be possible. An example of this situation is presented in Figure 6.2. While zones of greater and lesser ventilation rates can be determined, however, additional other testing will need to be performed if quantification of these local ventilation rates are desired. In the building investigation that generated Figure 6.2, the local ventilation rates were determined both by directly measuring the supply air and OA quantities with a flowhood and comparing supply air and return air tracer concentrations for the different AHUs serving this area.

Equipment Requirements

In order to perform the tracer decay method, the equipment required includes a source of the tracer gas, a means of collecting samples of the tracer in the room air, and a tracer gas analyzer. This analysis is usually performed using a gas chromatograph with an electron capture detector. An electron capture detector has a radioactive source that is a beta-emitter. Beta emissions consist of electrons. The other type of radioactive sources emit alpha particles. These are a concern with radon, and consist of two protons and two neutrons all together. Alpha particles are therefore the equivalent of the nucleus of a helium atom.

The electronics of the detector device convert this flow of electrons into a measurable current. The detector is normally surrounded by an inert gas such as argon, so that this "standing current" is maximized. When a sample of air is introduced, it goes through a packed column which separates the components in the air sample by their molecular size, the smaller molecules getting through more quickly. This separated sample then flows to the detector where the presence of elements which have an inherent affinity for electrons, such as the oxygen in the air or the fluorine in the SF_6; reduce the flow of electrons comprising the "standing current". The depletion occurs in an amount proportional to the number of molecules of that element which are present. The electronics of the device invert and zero out the standing current so that this decrease in standing current is displayed as an upward spike, proportional to the amount of material present. What needs to be remembered about this type of analytical equipment is that as the output signal increases, it can only do so as long as there is some electron flow still remaining. Eventually, the detector becomes saturated and the output signal is no longer linear with respect to the amount of tracer gas present in the sample. Therefore, the linear response range for the analytical equipment needs to be determined and periodically checked when this type of equipment is used.

Advantages and Disadvantages

Tracer decay measurements are a powerful tool for evaluating the performance of ventilation systems. One advantage of this approach is how quickly results can be obtained. One technique for estimating the time required to obtain useful data from a tracer test is to compute the halving time in tracer concentration. For a well-mixed space being provided with 1.0 ACH of OA, the halving time for the tracer concentration is 41.6 min. For a space with 0.5 ACH, the halving time is doubled to 83.2 min. Therefore, the time interval that the HVAC parameters need to be held constant during the test is in the range of 1 to 2 hours. Since the tracer will typically persist in the building for longer than this, then the HVAC parameters can be varied to increase the information obtained. After sufficient data has been collected with the OA dampers in one position, for instance, the position of these dampers could be changed and the impact of this change on the ventilation rate could then be quantified. This advantage contrasts favorably with the use of CO_2 measurements as a determination of ventilation rates. If one used CO_2 measurements to determine a ventilation rate of 1.0 ACH, for instance, it would take almost 4 hours for equilibrium conditions to be achieved. As discussed above, the achievement of equilibrium conditions and the maintenance of both constant HVAC parameters and population numbers are required to convert the peak CO_2 value to a ventilation rate.

Another requirement for the appropriate application of the tracer decay method is that the space being evaluated needs to be definable as a single zone and the decay must begin with a uniform tracer concentration throughout the space. It must be remembered that the results provided are the *average ventilation rate,* in ACH, for the entire zone of air movement and not just for a particular sampling location.

If the tracer decay measurements plotted on the semilog graph paper do yield a straight line with respect to time and the air change rate has remained constant over that time, then one can be confident that the data obtained are accurate within the assumptions necessary for the validity of the tracer dilution method. Some scatter of points is expected and the straight line may require a "best fit" approach. A minimum of three points over 1 hour should be used to determine this straight line. Once this line has been determined, any two points can be chosen with the coordinates (C_1, t_1) and (C_2, t_2), where C_i is the concentration at time i. The air change rate, I, can then be calculated as follows.

$$I = \frac{\left(\ln C_1 - \ln C_2\right)}{\left(t_2 - t_1\right)} \qquad (6\text{-}7)$$

This graphical method lends itself well to the field study of the data since it is easy to plot the log of the concentration as a function of time. It is less sensitive to errors in concentration than other methods. It has the further advantage that a graph provides a visual display of any departures from the straight line required by the exponential decay law. One limitation of both the tracer testing and the CO_2 approaches is that they cannot distinguish between OA arriving at the test location that is delivered by the ventilation system and that which is due to infiltration. If there is a need to distinguish between these two differing sources of OA, and there usually is since it is the performance of the ventilation system that is being evaluated, there is a third assessment technique available that is based specifically on the OA and SA quantities that are delivered to the area being evaluated. This latter approach therefore excludes the contribution of infiltration.

OUTDOOR AIR COMPONENT OF TOTAL DELIVERED SUPPLY AIR METHOD

The third approach that can be used to determine the quantity of outdoor air delivered to the building occupants relies on the measurement of both the total supply air (SA) quantity and a determination of the %OA in that supply air.

This is a basic approach which relies on two separate determinations. One step involves the measurement of the SA quantities being discharged from each of the supply diffusers serving the area of concern. The other step is a determination of the %OA in this SA based on the measurement of the outdoor air, return air (RA), and mixed air temperatures. Combining the two yields the amount of outdoor air delivered to this space.

Example 6.3. Consider a 10,000-ft² space with 70 people in it. This is a density of 7 people per 1000 ft². The balancing hood measurements of the 130 supply diffusers serving this space yield a total SA quantity to this space of 8400 cfm. The return air temperature to the AHU is 72°F, the outdoor air temperature is 32°F, and the mixed air temperature, based on the average of 12 measurements, is 66°F. Dividing the difference between the return and mixed air temperatures (6°) by the difference of the return and outdoor air temperatures (40°) yields a result of 15% outdoor air. Multiplying the total SA quantity of 8400 cfm by 0.15 then yields a delivered OA quantity of 1260 cfm. This OA quantity is equal to an average delivery of 18 cfm of outdoor air for the 70 people present in the building. In a situation like this, one should also check that the distribution by location for the delivery of the air corresponds to the distribution of the people.

COMPARISON OF OUTDOOR AIR QUANTITY ENTERING AHUs WITH OUTDOOR AIR QUANTITY DELIVERED TO BUILDING OCCUPANTS

In Chapter 5, the information presented focused on determining the quantity of outdoor air entering the HVAC system at the AHUs. This quantity can then be compared with the results obtained from the techniques discussed to this point in this chapter — the quantity of outdoor air determined to be actually reaching the building occupants. Three basic relationships can exist between these two quantities:

1. These two quantities of outdoor air are equal.
2. The quantity of outdoor air entering the AHU is less than that determined to being delivered to the building occupants.
3. The quantity of outdoor air entering the AHU is more than that determined to being delivered to the building occupants.

There are two controlling factors in each of these relationships: the amount of leakage from the SA ductwork to the return plenum, and the amount of infiltration of outdoor air directly from the outdoors. Therefore, the first condition, where the measured OA quantity entering the AHU and being delivered to the building occupants could indicate the situation where there is no leakage from the SA ductwork to the return plenum *and* there is no infiltration of outdoor into the occupied spaces. One could also conceive of a situation where this result could also be achieved when the amount of leakage of outdoor air from the supply ductwork was exactly equaled by the amount of infiltration of outdoor air. However, because the total volume of air into and leaving a given space must be equal over time, the above situation could only occur where the supply air was 100% outdoor air.

For the second condition, where the quantity of outdoor air entering the AHU is less than the quantity being delivered to the building occupants, the underlying situation necessary would have a considerable amount of outdoor air infiltrating into the occupied spaces. For this to be occurring, however, the location under evaluation would need to be operating at a negative pressure with respect to the outdoors. This condition should not be occurring in a correctly balanced building and would have been checked for, as discussed in Chapter 8.

For the remaining third, and most likely condition, where the OA quantity entering the AHU is more than that determined to being delivered to the building occupants, there can be several reasons for this situation to exist. The potential contributing factors include not only the existence of leakage from the SA ductwork to the return plenum, but also the mixing relationship between the supply air and the air already in the occupied space. For instance, a major portion of the supply air may merely short-circuit across the ceiling of the occupied space to the return grille, without actually mixing with the air at the breathing height of the occupants in this space. The terms "ventilation effectiveness" and/or "efficiency" attempt to evaluate this situation by providing an analytical approach to assessing the how ventilation air accomplishes the tasks of replacing the "old" room air with clean air and removing air contaminants from this space. The characterization of this aspect of ventilation rates is discussed in Chapter 7.

In addition to these two terms, let me also suggest two others — that of distribution integrity (DI) and distribution apportionment (DA). The DI term refers to the ability of the distribution components of the HVAC system to transport the supply air, after it leaves the AHU, and actually get it delivered to the occupied portions of the building. The distribution apportionment term, DA, refers to the design and operation of the ventilation in relation to the distribution of the people in the occupied spaces of the building.

In addition to the terms DI and DA, there are several other terms that are used to characterize the performance of ventilation systems from the perspective of the in-space mixing characteristics and airflow patterns. These other evaluation criteria include the terms ventilation efficiency, ventilation effectiveness, air-exchange efficiency, pollutant removal effectiveness, and age of the air. There has yet to be consistent agreement among the various investigators in this field as to the precise meaning and determination procedure for quantifying these values. In this book, the attempt is to provide some clarification as to the use of these terms based on what makes the most sense to me based on my own experience.

The first clarification is that both the DI and DA terms deal with the percentage of the outdoor air entering the AHU that gets delivered to the occupied spaces via the supply diffusers, while the other terms deal with what happens with this air volume after it leaves the supply registers. Based on this distinction, the boundary between the DI and the DA terms and the other evaluation criteria of the characteristics of the ventilation system is at a point just a few inches beyond the SA diffusers.

DISTRIBUTION INTEGRITY

As introduced in the discussion in Chapter 4, the OA quantity entering the HVAC equipment may not, and typically does not, represent the OA quantity that gets delivered to the building occupants. The calculation of the DI can therefore be written in the form presented in the following equation:

$$DI = \frac{\text{OA entering AHU} - \text{OA leaving supply registers}}{\text{OA entering AHU}} \times 100$$

Although the terms in this equation specify the use of the outdoor OA quantity (typically in cfm), the value for the SA quantity can be substituted since the OA quantity will remain a uniform percentage of the supply air in the distribution system. Applying this equation will yield values that can be as high as, but never exceed, 100% for those rare situations where there are no distribution losses. Actually measured values are typically around 85% or better, although I have evaluated buildings with values as low as 25%.

As indicated in the above equation, the magnitude of the distribution integrity can be determined merely by the difference between the OA quantity measured to be entering the AHU and the OA quantity measured to be delivered to the occupied space. From a practical standpoint, however, the actual determination of the quantity of outdoor air entering the AHU can be a difficult measurement to obtain. The difficulty lies not in the determination of the %OA, but in the ability to measure the volumetric flow in any of the three air streams — outdoor air, return air, or supply air. This difficulty is typically due the combination of the lack of appropriate locations for obtaining reliable flow measurements and the lack of access to straight lengths of ductwork. These conditions result from the severe space limitations for the mechanical equipment predicated by the maximization of rentable space. There are, however, other ways of assessing the DI. The importance of the DI of an HVAC system can be appreciated when, as part of an IAQ evaluation, it is determined that there is an inadequate supply of outdoor air being delivered to the building occupants. Before this situation can be corrected, it must first be determined why this is occurring. Is it because the OA quantity entering the AHU is inadequate to begin with, or perhaps this quantity is potentially adequate but the distribution losses are so great that not enough air, both outdoor air and supply air, is being delivered? The third possibility is, of course, that it is due to a combination of these two factors.

There are several reasons why the OA quantity entering the HVAC system does not get delivered to the building occupants. One major reason why these two quantities can differ can be due to the fact that a portion of the supply air is short-circuiting directly into the return plenum, due to leakage from its ductwork, and thereby preventing this air from reaching the people.

A feeling for the magnitude of the number of connections, and therefore potential leakage sites, that exist can be appreciated by listing the specifics of one building's distribution system. For this one HVAC system, which is serving an area of 23,000 ft^2, the distribution portion includes 10 reheat coil and VAV boxes with booster fans, 20 VAV boxes without fans, 144 slot diffusers, and 1860 linear feet of duct. For each metal-to-metal duct connection to be leak-tight, not only must each joint have had a caulking material applied, but this material also needed to have been permitted to cure for 24 hours before the system was pressurized since this ductwork shares the return plenum with other components of the building system, such as the structural, plumbing, and electrical systems. Access to all sides of the ductwork is not always easily achieved and therefore may not be adequately sealed. If the building has had areas renovated, or where changes in use have resulted in modifications to the distribution system, the potential for leakage sites is even greater because access within the plenum can be expected to be reduced during renovation as compared with during new construction. I have also observed supply duct openings that discharge directly into the return plenum.

In the definitions given above, the boundary defining the end of the distribution system can be described as a point just beyond the SA diffusers. The following example explains the reason for not merely saying the end-point of the distribution system is the diffuser itself.

Example 6.4. In one building, the short-circuiting was traced to the supply diffusers. These diffusers, of the troffer type, were discharging the supply air via a slot adjacent to the light fixtures. These diffusers, however, were unfortunately both an inexpensive brand and poorly installed. Our investigation determined that up to 50% of the supply air was being sucked directly into the return plenum at the point of discharge. This situation was aggravated, in terms of both the IAQ and the thermal comfort being provided to the building occupants, in that the thermal sensors were located in the return plenum as opposed to being in the occupied spaces. Therefore, when it was sufficiently overheated in the space that the sensors in the plenum finally called for cooling, the plenum was cooled off more quickly than the occupied space was, and the thermal sensors indicated that the cooling requirement had been met while the occupied spaces remained overheated.

DISTRIBUTION APPORTIONMENT

Another reason why an adequate OA quantity may not get delivered to the building occupants, despite a potentially adequate OA quantity entering the HVAC system, is because the supply air may be going to locations where there are no people, such as corridors and lobbies or exhaust systems.

Example 6.5. In one building most of the building's fresh air went to the basement where most of the people weren't. The specifics of this building are summarized in Table 6.7. One reason that the majority of the air was going to the basement was that there was 12,000 cfm of process exhaust leaving this location.

The most glaring fact to be garnered from a review of the data presented in Table 6.7 is the information for the basement that indicates that this location had only 3% of the people in the building, and yet received 52% of the outdoor air. In contrast, the first floor, with 33% of the people, received only 7% of the outdoor air.

Table 6.7.   Occupancy Loading and Outdoor Air Quantities

Location Floor	Occupancy Loading		Outdoor Air		
	People	%	cfm	%	cfm/Person
3	274	34	4,900	22	18
2	240	30	4,350	19	18
1	268	33	1,650	7	6
B	28	3	11,600	52	414
Totals	810	100	22,500	100	28

REFERENCES

1. American Society of Heating, Refrigerating, and Air-Conditioning Engineers, 1989. *Standard 62-1989: Ventilation for Acceptable Indoor Air Quality.* Atlanta, GA.
2. Persily, A. and W. S. Dols. 1990. "The Relation of CO_2 Concentration to Office Building Ventilation". *Air Change Rate and Airtightness in Buildings,* ASTM STP 1067. M. H. Sherman, Ed. American Society for Testing and Materials, Philadelphia. pp 77–92.
3. American Society of Heating, Refrigerating, and Air-Conditioning Engineers. *1989 Fundamentals Handbook.* Atlanta, GA. p. 23.2.
4. Balvanz, J., S.C. Bowman, A. Fobelets, T. Lee, K. Papamichael, and R. Yoder. 1982. *"Examination of Cabin Environment of Commercial Aircraft",* Iowa State University. 139 pp.
5. American Society for Testing and Materials. 1983. *"E741-83, Standard Test Method for Determining Air Leakage Rate by Tracer Dilution."* Philadelphia, PA.

CHAPTER 7

Ventilation Characterization

OVERVIEW

The term "ventilation characterization" deals with the airflow patterns that exist in the space being ventilated and the resulting ability of that airflow to dilute and remove air contaminants from that space. Chapters 5 and 6 focused on the evaluation of the quantity of air being delivered to the space. This chapter is concerned with the evaluation of what happens once this air is delivered to that space. There are four basic descriptions of airflow patterns that can exist in a given space; they are:

1. plug (or piston) flow
2. displacement flow
3. completely mixed
4. short circuiting

For the delivery of a given volume of air to an occupied space, the rate of dilution and removal of air contaminants from that space will vary as a function of the airflow patterns that exist in that space at that time. Figure 7.1 presents schematics of these four categories of airflow. Over the years, various methods have been proposed for evaluating the in-space characteristics of ventilation performance. This chapter first summarizes the various parameters involved and then provides a chronological review of some of the major efforts in this field that have attempted to define techniques for being able to evaluate and quantify the ventilation characteristics of a given space.

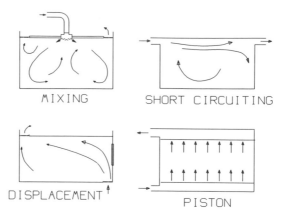

Figure 7.1   Four categories of air flow in rooms.

VARIABLES AFFECTING AIRFLOW PATTERNS

Differing airflow patterns within an enclosed space will vary in their ability to deliver supply air (SA) to the breathing zone of the occupants. The performance characteristics of a ventilation system, in a given space, will depend on the following variables:

1. The temperature of the delivered supply air
2. The temperature of the space
3. The type and positioning of supply and exhaust terminals
4. The location of objects in the space
5. The shape and size of the room
6. The presence or absence of heat sources in the space
7. The injection velocity of the supply air

The importance of some of these variables should be obvious, while the significance of others are discussed in the review of the selected references that follows.

A SUMMARY OF SELECTED REFERENCES

My files on the topic of ventilation efficiency and effectiveness go back 11 years. Since then, numerous papers have been published on this subject and more are still being published every year. The following is a chronological summary of papers I have selected to present as an overview of this material. Before the reader begins this section, let me first point out that the degree of control for the evaluation of performance characteristics of a ventilation system will be very different for a controlled research environmental chamber than for an actual occupied office building space.

Ventilation Efficiency: Sandberg and Svenson

In terms of defining the various terms in use for assessing the in-space ventilation characteristics, the oldest document in my files dates to 1981 and is the work of Sandberg and Svenson,[1] who were with the Heating and Ventilating Laboratory, Building Climatology and Installation Division, National Swedish Institute for Building Research, Gavle, Sweden.

The term they use to characterize the performance characteristics of the ventilation system is "ventilation efficiency." They define this term as the ability of the system to evacuate air contaminants from a particular source in a room. In this context, these air contaminants are to be interpreted in general terms to include such things as odors, radon, or excessive heat. In order to facilitate the computation of a value for this term, however, there needs to be an air contaminant that can be expressed in measurable units, such as ppm (parts per million) or ppb (parts per billion). Their definition of ventilation efficiency is based on two characteristics of a ventilation system:

1. The *relative ventilation efficiency* which expresses how the ventilation capability of the system varies between different places in a room
2. The *absolute ventilation efficiency* which expresses the ability of the ventilation system to reduce a pollution concentration in relation to the feasible theoretical maximum reduction

Sandberg and Svenson's computational approach for these values, however, is based on a comparison of the air contaminant concentrations in the supply air, the occupied zone, and the exhaust air, when there is both "an even distribution of pollution concentration in a room, and when there is a homogeneous source of pollution, i.e., the production of pollution in independent of time and equal everywhere in the whole room." While these conditions may be achievable in an environmental test facility, they are not practical in typical office environments.

According to their definitions of ventilation efficiency, η, summarized in Equation 7-1 and 7-2, is derived from a measured concentration of a component of the exhaust air, supply air, and working air. The working air refers to measurement obtained in the occupied zone of the space. Typically, a tracer gas is used as the representative component of the air. With these definitions, the relative ventilation efficiency is always positive and can be greater than 1, while the absolute ventilation efficiency is always less than 1 and can be negative.

$$\eta_{relative} = \frac{C_{exhaust} - C_{supply}}{C_{working} - C_{supply}} = 1 + \frac{C_{exhaust} - C_{working}}{C_{working} - C_{supply}} \qquad (7\text{-}1)$$

and

$$\eta_{absolute} = \frac{C(0) - C_{working}}{\Delta C_{max}} = 1 + \frac{C(0) - C_{working}}{C(0) - C_{supply}} \qquad (7\text{-}2)$$

where $C(0)$ = the initial concentration.

Among their findings, Sandberg and Svenson point out that the problem of defining a "local air change rate" from the slope of a tracer decay semilog plot is related to the fact that a room is generally part of an interlinked system. This means that changes in air contaminant concentration depend on the dilution process in other parts of the room, or zone for that matter. It is only under conditions of complete mixing that an unequivocal local air change rate can be defined from the slope.

For their particular investigation, Sandberg and Svenson's results also showed a heavy reduction in the relative ventilation efficiency with an increase in the supply air temperature. In the case tested, the relative ventilation efficiency at 3 ft off the floor (0.9 meters) decreased from 93% under isothermal conditions to 43% when the temperature of the air entering the room exceeded the temperature in the room by 4°C. The test chamber had dimensions of 11.8 × 13.8 × 8.86 ft high (3.6 × 4.2 × 2.7 meters), with the SA terminal mounted centrally in the ceiling and the exhaust air terminal on one wall, 8 in. (0.2 meters) below the ceiling. The nominal air change rate during this test was 4 ACH, which is indicative of a 100% outdoor air (%OA) condition. An aspect of this situation which points out the sensitivity of the results from this type of testing, reducing the slot opening of the SA terminal, led to an increase in the ventilation efficiency up to 85% when the temperature of the air entering the room exceeded the temperature in the room by 4°C. By adjusting the supply air terminal in this particular case, it was possible to improve ventilation efficiency by a factor of 2.

SA temperatures warmer than the space air temperature are typical for residential buildings and the perimeter of office buildings in climates with heating requirements. The interior of offices, however, rely on the delivery of air which is cooler than the in-space air temperature in order to achieve thermal equilibrium with the internal gains, that is, people, equipment, and lighting, that give off heat in that space.

Continuing with controlled laboratory investigations, there were two ventilation efficiency papers published in the 1982 special issue of *Environment International* devoted to indoor air pollution. Of particular note is one by Malmström and Ahlgren and another one by Skåret and Mathisen.

Ventilation Efficiency: Malmström and Ahlgren

The paper by Malmström and Ahlgren[2] was entitled "Efficient Ventilation in Office Rooms." The study was based on the use of a simple two-box model and the results obtained from tracer decay measurements where three different approaches to ventilation efficiency were compared. With respect to the tracer decay measurements, the authors point out that although the specifics of the transients vary, over time these plots tend to become parallel straight lines in the log-linear graph. The slope of these lines is $- n\varepsilon_I$, where n is the total air change rate (room volumes/time). That the decay lines become parallel over time indicates that the different zones exchange air between each other and that the rate of concentration decrease is governed by the zone with the lowest ventilation. They point out that the ε_I term has also been shown[3] to reflect the ratio of the transient parts of the tracer gas concentration in the exhaust air and the average concentration in the room. This ε_I term depends on the ventilation arrangement and the steady-state airflow in the room. If, in a given situation, the tracer decay lines on a log-linear plot become parallel at a time t_p after the start of the experiment, then another way of expressing the three ventilation efficiency terms, ε_I, ε_{II}, and ε_{III} is as follows:

$$\varepsilon_I = \frac{C_E - C_e}{C_R - C_r} \qquad (7\text{-}3)$$

$$\varepsilon_{II} = \frac{C_E - C_S}{C_W - C_S} \qquad (7\text{-}4)$$

$$\varepsilon_{III} = \frac{C_E - C_S}{C_R - C_S} \qquad (7\text{-}5)$$

where
C_E = the steady-state concentration in the exhaust air
C_e = the concentration in the exhaust air at time $t > t_p$
C_W = the steady-state concentration in the working zone
C_S = the concentration in the supply air (constant)
C_R = the steady-state mean concentration in the room
C_r = the mean concentration in the room at time $t > t_p$

All three of these terms, ε_I, ε_{II}, and ε_{III}, can be used as measures of "ventilation efficiency." Their discussion is based on the conclusions derived from a two-box model. The authors point out, however, that for actual rooms, important information about the performance of the ventilation system may be achieved by tracer gas measurements performed both with artificial mixing in the room (in order to establish the total air exchange rate n) and then followed by a test without artificial mixing. The result, however, will only be valid for the room airflow during the experiment. These authors comment that due to "the difficulty of achieving a stable airflow in an office room, this fact implies that attempts to attribute a certain value of 'ventilation efficiency' to an installation or a ventilation system may prove to be of doubtful practical value." The test does, however, have the potential to indicate to an experienced ventilation engineer a great deal about the tendencies for the air flows in the room and how well the type and location of air terminal devices have been chosen.

Ventilation Efficiency: Skåret and Mathisen

The second paper in the 1982 *Environment International* document, by Skåret and Mathisen,[4] was entitled "Ventilation Efficiency." In this paper, the authors adopt the ratio of $-\lambda_1/n$ as a transient ventilation, ε_I:

$$\varepsilon_I = -\frac{\lambda_1}{n} \qquad (7\text{-}6)$$

In addition to this transient value for ventilation efficiency, they also discuss a steady-state value, ε_{II}, that is calculated from the ratio between the concentration in the exhaust air, C_E, and the mean concentration for the room in question, C_R[3]. They report that these different methods of measuring and defining ventilation efficiency give essential differences in the results. The ε_I value gives a measure of the average speed at which a ventilating system dilutes contaminants brought into a room compared to the rate achieved dilution due to the complete mixing case. The second definition, ε_{II}, the ratio between the concentration in the exhaust air and the working zone at steady state, gives a measure of the ability of the ventilating system to remove contaminants from the zone of occupation.

One of their conclusions is that for achieving ventilation in conjunction with heating, the best system has the air supply located just beneath the ceiling and the air exhaust near the floor. For achieving ventilation while providing cooling, they suggest that a system with air supply near the floor and exhaust near the ceiling would be the best approach. That is, "diagonal schemes" seem to be the most efficient, and more efficient than having complete mixing. Remembering that the perfect mixing situation usually gets assigned a value of unity, then improvements over this approach earn values greater than 1 for their ventilation efficiency.

Another of their findings is that the difference in concentration between various locations for steady state and transient state, as obtained from tracer decay tests, are not equal.

One of the next significant events in the evolution of ventilation efficiency approaches was the 3rd International Conference on Indoor Air Quality and Climate, held in Stockholm from August 20 to 24, 1984, where three papers were presented that related to ventilation characterization.

Ventilation Effectiveness: Skåret

One of these was a paper by Skåret,[5] who was reporting on the results of tests carried out at the Lawrence Berkeley Laboratory in California during his sabbatical there from the Norwegian Institute of Technology during the 1982–1983 academic year. One of the significant things about this paper is that it introduces the concept of the age of the air as another technique for characterizing the performance of ventilation systems. Skåret stresses that the age of air analyzing techniques is an excellent tool for assessing ventilation effectiveness. He says that it is important to differentiate between air exchange effectiveness and contaminant removal effectiveness. This difference is a result of the conclusion that the distribution of air and contaminants are different in a ventilated space. From the perspective of the replacement of air in the room, merely relying on the air exchange rate (i.e., the ventilation airflow rate divided by the room volume) is by no means satisfactory to express either air renewal or air quality. The inverse of this term, the turnover time for the ventilation airflow through the room, τ_n, is calculated by dividing the effective room volume by the ventilation airflow rate.

Example 7.1. Assume a space with the dimensions of $20 \times 50 \times 10$ ft high that has six occupants and a supply air rate of 850 cfm that has 15% OA This results in a ventilation airflow rate of 127.5 cfm. Estimating the effective volume as 85% of the total gross volume of 10,000 ft^3 yields a value of 8500 ft^3. Dividing the ventilation airflow rate (127.5 cfm) by the room volume (8500 ft^3) yields a ventilation rate of 0.015 AC/min, or 0.9 ACH. Taking the inverse of this value yields the turnover time, τ_n, for air flowing through the room as 1.111 hours, or 66.667 min.

The turnover time (average air exchange time) for the total volume of air in the room is twice the average age for the air in the room, taken as the time elapsed after the air entered the room. Now the airflow patterns become important and, consequently, this time quantity differs significantly between different ventilation systems.

Another way of looking at this is that during the process of supplying air to a room, more or less air is lost through the exhaust before all the air in the room is replaced with new air. Therefore, for any arbitrary time zero, the turnover time for the air in the room is now the average time it takes for every molecule of the air that was in the room at time zero to be replaced by a new molecule.

Comparing the turnover time for the air in the room to the average air exchange time, one finds that only for the special case of air flowing through the space as a piston will these two times equal each over. For conditions other than plug flow, the average exchange time for air *in the room* will be greater than that for the turnover time for the air *entering and leaving the room*. For the complete mixing case, the average exchange time for air *in the room* will be twice that for the turnover time for the air *entering and leaving the room*.

Example 7.2. With plug flow through the space, a ventilation rate of 1 ACH, and an interval of 1 hour, all of the molecules of air in the room will have been replaced. With perfect mixing, the same 1-ACH ventilation rate, and 1-hour interval, 36.8% of the air molecules will still remain. This is the exponential decay situation, where the amount leaving the space is proportional to the concentration remaining in the space. This relationship was discussed previously in Chapter 5, and Equation 5-4 can be used to calculate the amount tracer or air molecules remaining.

In addition to assessing the renewal rate for the air in the room, the other side of the equation in evaluating indoor air quality (IAQ) is the rate at which air contaminants are removed from the space. The author also points out that the turnover time for the air contaminants are, contrary to situation for the ventilation air, dependent on the flow patterns for the total system. Displacement flow of ventilation air promotes short-circuiting of contaminants, which is a desirable goal. However, the contaminant source may have a thermal buoyancy or momentum flux of its own, which can either work with or against the rooms' flow patterns. This uncertainty makes it difficult to predict turnover times for the contaminant flow based on age determination for the air in the room. Local concentrations may vary considerably, leading to the situation where air quality can be determined only from measurements at all locations in the room. However, the more uniformly distributed and passive the sources are, the better the correlation is between air quality and average age of air in the room. *On the other hand, for sources that are known to have an upward momentum, typical for sources that are warmer than their surroundings (such as people, equipment, or cigarettes), the most effective displacement flow would therefore also upward through the space.*

They summarize by stating that the air exchange effectiveness can be expressed as the ratio between the turnover time for the ventilation air and the turnover time for the air in the room:

$$\bar{\varepsilon}_e = -\frac{\tau_n}{2\langle \tau_i \rangle} \tag{7-7}$$

where $\tau_n = V/V' =$ turnover time for the ventilation airflow.

Since the inverse of the slope of the last part of the decay curve, λ_e, after the tracer release has terminated is an indicator of the numerical value of the quantity $< \tau_i >$, ε_e may therefore be roughly expressed as:

$$\bar{\varepsilon}_e \cong -\frac{1}{2}\tau_n \lambda_e \tag{7-8}$$

Ventilation Efficiency: Janssen

Another paper presented at the 3rd International Conference on Indoor Air Quality and Climate was by Janssen,[6] entitled "Ventilation Stratification and Air Mixing." This paper continued in the trend of using a two-chamber model to develop ventilation efficiency equations. His conclusion is that tracer gas measurements from just the occupied zone can be used to calculate values for ventilation efficiency and stratification or mixing factors. There are three requirements for this procedure:

1. the ventilation is substantially greater than the exfiltration
2. there is poor mixing
3. there is a measurable difference between the initial tracer decay rate, I_0,
   the final rate, I_∞, in the room

If so, then the following equation can be used:

$$\eta \approx \frac{I_0 - I_\infty}{I_0}$$ (7-9)

Ventilation Efficiency: Sandberg and Sjöberg

The third paper presented at the 3rd International Conference on Indoor Air Quality and Climate that pertained to ventilation efficiency was by Sandberg and Sjöberg[7] and was entitled "A Comparative Study of the Performance of General Ventilation Systems in Evacuating Contaminants." This paper continued with the ideas brought forth by Sandberg concerning the usefulness of age of air calculations. Specifically, they compare three ventilation schemes with the following relationships between the supply air terminal and the exhaust air terminal: ceiling-ceiling, ceiling-floor, and floor-ceiling. Their conclusion was that the parallel-flow systems, where the air and the contaminants move in the same direction (i.e., upward through the space), give rise to the lowest average room concentrations.

Since piston flow gives rise to the fastest exchange of the air in the room, the authors have decided that this condition should be used as the reference case. This differs from other approaches, where the complete mixing case has been selected as the reference case. This selection leads to the following definition for *air exchange efficiency.*

$$\varepsilon_a = \frac{\dfrac{\tau_n}{2}}{\langle \tau \rangle} \times 100$$ (7-10)

where

ε_a = the air-exchange efficiency %
$\langle \tau \rangle$ = mean age of the air in the space with volume V
τ_n = nominal time constant of the ventilated system, V/Q
Q = the total outdoor airflow into the space

Example 7.3. Assume again a space with the dimensions of 20 × 50 × 10 ft high that has 6 occupants and a SA rate of 850 cfm that has 15% OA. This results in a ventilation airflow rate of 127.5 cfm. Estimating the effective volume as 85%

of the total gross volume of 10,000 ft³, yields a value of 8500 ft³. Dividing the ventilation airflow rate (127.5 cfm) by the room volume (8500 ft³) yields a ventilation rate of 0.015 AC/min, or 0.9 ACH. Taking the inverse of this value yields the turnover time, τ_n, for air flowing through the room as 1.111 hours, or 66.667 min. If this space is perfectly mixed, then all the local ages have the same value throughout the space, which is equal to the age of the air leaving the space. Therefore, $<\tau> = \tau_n$. Inserting these values in Equation 7-10 yields the following relationship.

$$\varepsilon_a = \frac{\dfrac{1.11\ h}{2}}{1.11\ h} \times 100 = 50\% \tag{7-11}$$

Adopting the piston-flow case as the reference case (i.e., efficiency of 100%) makes a lot more sense to me than the previous situations where the reference case of perfect mixing because then displacement conditions lead to efficiencies in excess of 100%.

Adopting the definition of air exchange efficiency, as presented in Equation 7-10, can then become the basis for evaluating and comparing airflow patterns in various spaces and configurations of supply and return terminals and the other variables. The basic term in the numerator, τ_n, should be straightforward at this point; that is, the volume of the space divided by the total outdoor airflow into that space. The challenge in using this equation then becomes the calculation of the $< \tau >$ term, the mean age of the air in the space. Two papers presented at the ASHRAE IAQ '86 Conference in Atlanta, Georgia deal with this very issue.

Ventilation Characteristics: Seppänen

One of these papers was authored by Seppänen,[7] who points out that there are several ways to describe ventilation characteristics. He distinguishes between the two basic tasks of ventilation, the replacing of "old polluted" room air with clean air and removal of pollutants from the room and preventing them from spreading to the occupied zones in the space. The first task is described by the calculation of the air exchange efficiency that has already been described. Since airflow rates are usually specified in standards, assuming complete mixing in the space, the actual achievement of the desired level of "air quality" will vary as a function of the airflow pattern: a greater airflow requirement if the air distribution is short-circuiting and a smaller airflow if there is a displacement flow pattern. The relationship can be expressed as:

$$q_{design} = 0.5 \frac{q_{std}}{\varepsilon_a} \times 100 \tag{7-12}$$

where

q_{design} = the airflow to be used in the actual design of the system with an air exchange
efficiency of ε_a

q_{std} = airflow from the standards, assuming complete mixing

If an air contaminant is uniformly distributed within the room air, then the air exchange efficiency will also provide an assessment of the pollutant removal effectiveness. However, since the concentration of air contaminants may not be uniformly distributed, the issue of pollutant removal effectiveness needs to be addressed separately. The author therefore suggests that a criteria for this task be defined as the ratio of pollutant (or tracer) concentration in the exhaust air to the concentration in the room air during steady-state conditions:

$$\epsilon_c = \frac{C_e}{C_r} \qquad (7\text{-}13)$$

where
ϵ_c = pollutant removal effectiveness
C_e = concentration in the exhaust air
C_r = concentration in the room air

Now the value for room air may be described in several ways. It can be the average in the room air or it may be limited to the zone of occupation or only to a certain point in the room.

With the complete mixing condition achieved, the pollutant removal effectiveness is equal to 1. With localized pollutant sources and local exhaust systems, removal effectiveness can be higher than 1.

Ventilation effectiveness is usually measured by the performance of tracer testing. Several methods can be used, but is typically the decay method. Starting from the achievement of a uniform distribution of the tracer gas throughout the space, the decay in concentration is measured at various locations. This permits the air exchange indices to be assessed. The age of the air at a certain point, t_p, is evaluated by integrating the area below the concentration decay curve down to the time axis and dividing it by the concentration at the beginning of the decay when the distribution was uniform. This relationship can be expressed as:

$$\tau_p = \frac{\int_0^\infty C_p(t)dt}{C_0} \qquad (7\text{-}14)$$

If the tracer is also measured in the exhaust duct, or ducts, the nominal time constant, τ_n, of the ventilation is obtained and the nominal air change rate, in ACH, is equal to $1/\tau_n$. As stated previously, if complete mixing is occurring, this nominal time constant of the decay is measured during complete mixing, τ_n, is equal to the average age of the exhaust air, or the air at any point in the room. If complete mixing is not occurring, then the average age of the air in the room, $< \tau >$, can be calculated from the measurements in the exhaust duct:

$$\langle \tau \rangle = \frac{\int_0^\infty t\, C_e(t)\, dt}{\int_0^\infty C_e(t)\, dt} \qquad (7\text{-}15)$$

According to the author, the average age defined from exhaust air measurements describes the operation of air distribution in general, but does not give any information on ventilation or air quality. The evaluation of ventilation and air quality requires the measurement and calculation of the local age of the air from Equation 7-15. He reports that, based on field measurements, there seems to be more uncertainty in the average age of the air as defined from exhaust ducts than from air that is measured locally in the room. Usually, after the mixing of the tracer gas, the decay is different at various points even if the slopes of the decay curves in logarithmic scale are the same in the steady state of the decay process. The difference in the initial decay transients indicates the differences in the local ages of the air.

There are still several practical difficulties in applying this method to actual office buildings. First, there is the difficulty in obtaining an initially uniform distribution of tracer gas at the start of the process. In Equation 7-15, the term C_0 needs to correspond to this uniform initial value. The value for this term can be estimated from the average of all of the initial points; however, there will be an error introduced if the differences between the points is large. Another difficulty is attempting to obtain tracer measurements in the exhaust duct which arises if the zone under evaluation is smaller than the zone served by the AHU. In this situation, the return air (RA) tracer concentration at the AHU will not necessarily represent that for the space being evaluated, and the room being evaluated may have too many exhaust terminals to be sampled from a practical standpoint.

In summary then, depending on how the delivered air moves through the occupied space, the delivery of identical quantities of outdoor air can therefore achieve tremendous differences in the rate of dilution and removal of air contaminants in that space. In addition to the ventilation rate, there are additional factors of the performance of the HVAC system that need to be considered. These include the air exchange efficiency, the age of the air, and the pollutant removal effectiveness. These factors can be combined to characterize the performance of ventilation systems.

Ventilation Effectiveness: Persily

As mentioned earlier in this chapter, there is a big difference between performing age of air measurements in a single laboratory chamber and in a large office building. A paper by Persily,[8] entitled "Ventilation Effectiveness Measurements in an Office Building" was also presented at the ASHRAE IAQ '86 Conference. This paper discusses both the tracer "build-up" and "decay" methods for determining ages of air, although the definition he presents goes back to a mean air exchange effectiveness defined by:

$$n = \frac{t_n}{\langle t \rangle}$$
(7-16)

where perfect mixing, at $n = 1$, is the reference case. This can be contrasted with the previous example where piston flow is the reference case.

The practical considerations in applying age of air calculations in office buildings include the fact that there may be air intake into the space due to means other than via the mechanical ventilation system. If this is occurring, it would eliminate the uniformity of the equilibrium concentration throughout the space for the build-up procedure. Such intake into the space includes both uncontrolled air leakage through the building envelope (infiltration) and airflow from adjoining spaces within the building. Another problem with the build-up procedure is the fact that most buildings are served by more than one air handler. These multiple air handlers generally serve zones that often communicate with one another in terms of airflow. In order to conduct the measurements, one would need to inject the tracer gas into each AHU such that the same tracer gas concentration would exist in the supply air of all of the systems. Such an injection scheme is practically impossible without precise knowledge of the supply airflow rates of the various AHUs and the air exchange rates of the corresponding zones.

The tracer decay procedure, however, also has practical limitations due to the physical complexities of office buildings. The theory of age distributions — and most of the laboratory measurements — considers the ventilated space as a single zone with a small number of well-defined supply and exhaust (return) vents. The space is also assumed to have no airflow into or out of the space except through the mechanical ventilation system. For typical office spaces, with ducted supplies and plenum returns, there are several other difficulties in applying the age of air measuring techniques.

Remembering that the measurements of the average age of the air in the space, $< t >$ (or $< \tau >$ depending on the author), require the monitoring of the exhaust concentration, then if one is attempting to determine the value of $< \tau >$, which is part of a multi-room zone, then the exhaust concentration will be difficult to define. Air will be leaving the room through exfiltration, airflow to adjoining rooms, and airflow into the RA system, often through more than one outlet. The tracer gas concentrations of the air leaving the room at each of the various points can be quite different. While one may be able to monitor the exhaust concentration in the return vents, it will be questionable as to whether this is truly representative of the exhaust concentration from the room.

One can attempt to determine $< t >$ for the entire zone by monitoring the exhaust concentration at the connection between the RA plenum and the RA shaft.

However, there will be some uncertainty if the exhaust concentration at this location actually provides an indication of the ventilation effectiveness throughout the entire space. In addition, there may be outside air infiltrating into this depressurized plenum space, and because of this airflow, the results may not reflect the conditions with the zone.

The existence of recirculation of RA is another issue in these measurements. When there is recirculation of RA, the measurements can still be conducted, but the results will reflect both the air movement patterns within the space and the mixing due to recirculation. Another problem is the fact that the amount of recirculation and OA intake may change during a test as the control system adjusts damper positions. In variable air volume systems, the total airflow rate through the system may also be changing as well. Again, the tests can be conducted, but the results may be biased by the changing conditions and it may be very difficult to obtain identical test conditions for validating the repeatability of the tests.

The tracer decay method may be difficult to use in buildings that have multiple AHUs. One challenge would be to be able to achieve the desired initial condition of a uniform tracer gas concentration throughout the building. Some success can be achieved by estimating the appropriate amounts of tracer to be injected into each system as a function of the air volume involved. After waiting for equilibrium to be achieved based on these initial injection rates, these injection rates can then be adjusted based on the differences in the tracer gas concentrations among the zones. One may use the RA concentrations for each zone as an indication of the concentration in that zone, although the existence of outside air leakage into the RA plenum means this concentration will not be a strictly accurate indication of the zone's concentration.

Persily[8] points out in his conclusion that the measuring procedures are involved and still under development, but the techniques are being continually refined. His actual building results indicate apparently good mixing of the air on a whole building scale; however, there was some evidence of local stagnant zones in the workspace. These apparently contradictory results point out the importance of distinguishing between the micro and macro aspects of building ventilation performance, and the difficulty in extracting information about the micro environment from the macro-based tests.

Tracer Decay Testing: Alevantis and Hayward

The question of how uniform the initial tracer distribution needs to be in order for age of air calculations to be meaningful was discussed at INDOOR AIR '90, the 5th International Conference on Indoor Air Quality and Climate in Toronto, Canada by Alevantis and Hayward[9] of the California Indoor Air Quality Program. Their paper entitled, "The Feasibility of Achieving Necessary Initial Mixing When Using Tracer Gas Decays for Ventilation Measurements," examined the degree of initial mixing of tracer gas with building air for different buildings. Recognizing this need to have the tracer concentration initially well mixed throughout a building, in order for ventilation efficiencies to be calculable, the authors state that if the tracer is initially incompletely mixed, the decay curve slopes of the tracer concentrations at the various locations eventually become equal, indicating only the overall ventilation rate. In a previous paper, Turk et al.[10] claim that "satisfactory" mixing of the tracer gas can be assumed when concentrations at the sampling sites in the building are within 10% of one another. However, the scientific rationale for this number is unsubstantiated. Further, Turk et al. reported that in the majority of the 38 commercial buildings they studied using the tracer decay technique, "satisfactory" mixing could not be achieved, particularly in those with multiple ventilation systems. Alevantis and Hayward performed tracer decay tests in four office buildings that were operated at both low and high air change rates with a computer-controlled, multi-location sampling technique. Their results indicated that the tracer gas mixed well with building air at low ventilation rates (0.5 ACH), but that at high ventilation rates (1.0 ACH or above), good initial tracer gas mixing conditions were not achieved on a consistent basis.

These results suggest several points for discussion:

1. The increased uniformity of initial tracer concentration achieved at low ventilation rates is due to the increased amount of recirculation occurring in these buildings.
2. This presents a kind of *Catch 22* because, at these low ventilation rates, the recirculation (and therefore redistribution) of tracer complicates the interpretation of the results.
3. This suggests that one way to proceed is to achieve a high degree of initial tracer uniformity by keeping the OA dampers closed during the tracer build-up, but once this uniformity is achieved, the OA dampers would be opened to their maximum to permit the evaluation of local differences in ventilation efficiency.

Ventilation Efficiency: Outdoor Air Supply Index

In recognition of the problems associated with applying ventilation efficiency calculations to office buildings, another approach called the Outdoor Air Supply Index (OASI) has been suggested by Farant, Greaves, and Robb.[11] This method is based on the constant release of tracer gas into the outdoor air into the ventilation system. The procedure also involves adjusting the ventilation system to a 100% OA condition. This requirement eliminates the redistribution of tracer via the recirculated air and facilitates the rapid achievement of steady-state conditions. Once steady-state conditions have been achieved which, as discussed in Chapter 5, will vary as a function of the ventilation rate, measurements of tracer concentrations are then obtained both at the breathing zone level in the space and for the supply air. The portion of the outdoor air delivered to the breathing zone is then estimated as follows:

$$\text{OASI } (\%) = \frac{\text{Tracer concentration, breathing zone}}{\text{Tracer concentration, supply air}} \times 100$$

The authors point out that this method provides results similar to those obtained with the methods of ventilation efficiency and effectiveness in buildings or in experimental setups. Besides not requiring complex equations and being easier to use in actual buildings, this estimate of the portion of the outdoor air delivered to individual locations (typically workstations) takes into account most of the variables that can influence this value. The only known exception is the building envelope leakage, a shortcoming that this method shares with the other methods.

Ventilation Efficiency: Farant, Nguyen, Leduc, and Auger

In terms of the various contributing factors that can affect ventilation efficiency, studies conducted in a test chamber and in two buildings by Farant, Nguyen, Leduc, and Auger[12] have shown that the amount of outdoor air a person receives in an office building can be affected by "the temperature of the air supplied to the room, the type of diffuser, and the type, size, and proximity of the air return inlet." They also state that, "the type of office selected, its design and layout, could also have a marked effect."

The reader interested in delving more deeply into the issue of ventilation effectiveness is encouraged to follow the development of *ASHRAE's Standard Project Committee (SPC) 129P: Ventilation Effectiveness*.

CURRENT AND FUTURE EFFORTS

While the methods to evaluate the ventilation characteristics continue to evolve and be refined, the lessons already learned from theoretical models and laboratory studies can be put to use in both the design of new facilities and in qualitatively assessing existing situations. With respect to new designs, the desired airflow patterns are already established, what remains are the technical challenges to successfully implement them. One commercial approach that is able to provide the upward flow through the occupied spaces is the Underfloor Air System marketed by Krantz of West Germany. This approach needs to be the design intent from the beginning since, as can be seen in Figure 7.2, there needs to be a raised floor to permit the distribution of the supply air. The supply air itself is filtered and thermally conditioned the same as with overhead ducted supply and plenum return distribution systems. Once the building has been designed for the under-floor plenum, the next biggest challenge with this type of system is the actual delivery of the air without causing drafts because in the interior zones of office building this supply air will be cooler than the air in the space. The Krantz solution to this potential problem involved their floor mounted Twist™ outlets designed to induce rapid mixing with the air around them. They are apparently concerned about the potential for drafts because they recommend minimum distances be maintained between the SA outlet and any occupant in a static

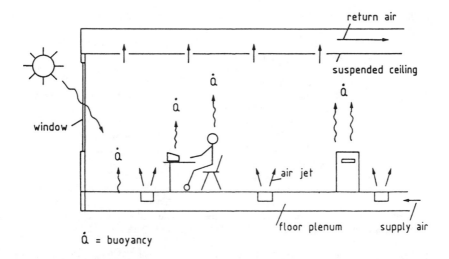

Figure 7.2   Sketch of under-floor air system.

position for prolonged periods of time. For their KB 150 floor outlet, the minimum distance is 2.5 to 3.3 ft, while for the KB 200, the minimum distance is 3.3 to 5 ft.[13] Worldwide, this system has been installed in 64 different buildings and has clear ventilation advantages over typical geometries where both the supply and return registers are located in the ceiling.

An alternative approach for the delivery of air from below the floor is the Filtered Air Control Technology (F.A.C.T.) which is an upflow displacement system that introduces the supply air through a perforated floor covered with commercial-grade carpet. This system was developed by research engineers at Philip Morris and they have completed one installation at the Benedum Center for Performing Arts in Pittsburgh.

These installations represent the future direction of the industry because of their ability to achieve either comparable ventilation with less energy consumption, or improved ventilation and therefore the potential for better IAQ, with the same consumption of energy as traditional ceiling-installed systems.

REFERENCES

1. Sandberg, M. and A. Svensson. 1981. "Measurements of Ventilation Efficiency by Using the Tracer Gas Technique." *Building Services Engineering Research and Technology,* 2:3 pp. 119–126.
2. Malmström, T.-G. and A. Ahlgren. 1982. "Efficient Ventilation in Office Rooms." *Environment International,* 8:401–408.
3. Malmström, T.-G. and J. Östrøm. "Något om lokal ventilasjonseffektivitet." A4-Series No. 47, Division for Heating and Ventilating, Royal Institute of Technology, Stockholm.

4. Skåret, E. and H. M. Mathisen. 1982. "Ventilation Efficiency." *Environment International,* 8:473–481.

5. Skåret, E. 1984. "Contaminant Removal Performance in Terms of Ventilation Effectiveness." *Proc. 3rd Int. Conf. Indoor Air Quality and Climate.* Stockholm. pp. 15–22.

6. Janssen, J. 1984. "Ventilation Stratification and Air Mixing."*Proc. 3rd Int. Conf. Indoor Air Quality and Climate.* Stockholm. pp. 43–48.

7. Seppänen, O. 1986. "Ventilation Efficiency in Practice." *IAQ '86: Managing Indoor Air for Health and Energy Conservation.* ASHRAE. pp. 559–567.

8. Persily, A. 1986. "Ventilation Effectiveness Measurements in an Office Building." *IAQ '86: Managing Indoor Air for Health and Energy Conservation.* ASHRAE. pp. 548–558.

9. Alevantis, L. E. and S. B. Hayward. 1990. "The Feasibility of Achieving Necessary Initial Mixing When Using Tracer Gas Decays for Ventilation Measurements." *Indoor Air '90: Proc. 5th Int. Conf. Indoor Air Quality and Climate.* Toronto, Canada. 4:349–354.

10. Turk, B. H., J. T. Brown, K. Geisling-Sobotka, D. A. Froehlich, D. T. Grimsrud, J. Harrison, J. F. Koonce, R. J. Prill, and K. L. Revzan. 1987. "Indoor Air Quality and Ventilation Measurements in 38 Pacific Northwest Commercial Buildings — Volume 1: Measurement Results and Interpretation." Lawrence Berkeley Laboratory Report, LBL-22315, University of California, Berkeley.

11. Farant, J. P., D. Greaves, and R. Robb. 1991. "Measurement and Impact of Outdoor Air Supplied to Individual Office Building Occupants on Indoor Air Quality." *Am. Ind. Hyg. Assoc. J.,* 52:387–392.

12. Farant, J. P., V. H. Nguyen, J. Leduc, and M. Auger. 1991. "Impact of Office Design and Layout on the Effectiveness of Ventilation Provided to Individual Workstations in Office Buildings." *IAQ '91: Healthy Buildings.* ASHRAE. pp. 8–13.

13. Sodec, I. F. and R. Craig. "Underfloor Air Supply System, Guidelines for the Mechanical Engineer." Report No. 3787 H. Krantz GmbH & Co.

CHAPTER 8

Air Movement Pathways and Pressure Relationships

OVERVIEW

Understanding the pressure relationships that exist in a building are an important part of an indoor air quality (IAQ) evaluation because it is pressure differences that provide the driving forces to transport air from one location to another. As discussed in Chapter 1, two of the four basic factors needed to cause an IAQ problem include a pathway for the transport of air contaminants and a driving force to transport these air contaminants from its source to the person or persons affected. This chapter deals with those two issues — the pathway and the driving force.

Therefore, understanding pressure relationships and the air movement patterns in the building that result from their existence are important because they can be responsible for the introduction or transport of air contaminants into or through the building to the occupied areas. The investigation of pathways of air movement are particularly important in responding to complaints about the presence of odors in the occupied spaces.

The identification of pathways begins by understanding the basic forces that exist within and around buildings. This knowledge guides the investigator to understand where to look for clues. The investigator then needs to be able to form a hypothesis to explain the known facts. The process can then continue with testing to confirm or refute this hypothesis, typically with the aid of tracer release testing.

RESPONDING TO ODOR COMPLAINTS

In responding to complaints about the presence of odors, there are other steps that should be taken in addition to those steps just listed. For one thing, an inquiry should be made to the people responsible for the management of the building to make sure that there is a mechanism in place so that any and all complaints are being taken courteously and are being recorded in a systematic fashion. This is important because it is frustrating enough for the occupants to feel that they are being exposed to air contaminants, without also feeling that their complaints are not being taken seriously.

These reports should be logged in at one location, if possible, and should include the location of the complaint in the building, the nature of the complaint, the date, the time, plus a notation on the outdoor weather conditions (temperature, wind speed, and direction), and any unusual occurrences. This information is important because each item is a potential clue as to the nature of the air contaminants and/or where they are coming from.

Example 8.1. In one problem building, odor complaints were traced to the introduction of sewer gas odors which, in turn, were related to the variation in the height of the tides. When the tides went out, a water seal was eliminated, and sewer odors found their way into the building. This was (of course) a coastal location.

Similarly, complaints can be related to certain wind directions and/or speeds if reentrainment is the transport mechanism and the air movement patterns created by the interaction of the wind and the building only foster this transport some of the time. The reentrainment of odors can involve sources such as kitchen exhausts, laboratory fume hood exhausts, or odors from rooftop plumbing vents. On the other hand, if the complaints correlate with a certain activity no matter what the wind direction and speed, then the pathway involved will typically involve a more direct route than the reentrainment of exhaust plumes back into air intakes. This more direct route will typically involve some pathway internal to the building.

A pathway is more than just a leakage site in the building; there also needs to be a mechanism to transport air from one place to another. It is the existence of building-related pressure relationships that govern the establishment of air movement pathways in that building.

When responding to odor complaints, it should be remembered that as part of the metabolism microorganisms, organic chemicals are generated and released. Some of these chemical compounds have significant odor potentials. Other sources of odors to be considered include specific chemicals associated with activities occurring at the building, such as diesel trucks at the loading dock or cooking fumes coming from the kitchen.

HOW AIR MOVES THROUGH BUILDINGS

With respect to the discussion of building pressure relationships and their associated air movement patterns, there should be an understanding of how air typically moves through a structure. Once it is understood how air typically moves through structures, the specific analytical methods for documenting their existence, identifying their specific routes, and even quantifying the dilution occurring along these pathways can be undertaken. In responding to odor complaints, the better one understands exactly how the air is transported from one location to another, the more effective the mitigation efforts can be to eliminate this undesirable transport of air. In many cases, one cannot eliminate the driving forces; thus the only option is to attempt to eliminate the specific pathway involved and make sure that no new pathways are permitted to be established elsewhere. This is because the driving force is typically due to the stack effect operating in the building, and this driving force cannot be eliminated during the heating season. In these situations, since the underlying cause cannot be eliminated, the options are limited to eliminating the pathways, relocating or eliminating the source, or relocating the people or the activity that is at the receiving end of the transported odors. In other cases, however, the driving force is related to the operation of the HVAC system, and this can sometimes be corrected.

BUILDING-RELATED PRESSURE RELATIONSHIPS

One of the major driving forces for the establishment of pressure differences in buildings is the temperature difference that can exist between indoors and outdoors. As air is heated indoors, it expands and becomes less dense and is consequently more buoyant than the colder outdoor air surrounding the building. When the indoor air is warmer than the outdoor air, it rises in the building and attempts to leak to the outdoors at the top of the structure. This buoyancy difference creates an outward pressure at the top of the building. When this air can escape to the outdoors, its departure creates a negative pressure in the lower levels of the structure which can then draw in colder outdoor air near the base of the building to replace the heated air which escaped. The driving force for this tendency for heated air to move up through the building, escaping at the top, and drawing in replacement air at the bottom is called the thermal "stack effect." The tighter the building shell, the higher the pressure that can build up at the top of the building, and the lower the resulting airflow volume.

One estimate of the magnitude of this pressure difference, in inches of water column (in., w.c.), due to this stack effect can be computed from the following equation from the *ASHRAE Fundamentals:*[1]

$$\Delta p_s = C_2\left(\rho_o - \rho_i\right) g \left(h - h_{NPL}\right) \tag{8-1}$$

This equation can be rewritten, keying the change in density to the temperature difference:

$$\Delta p_s = C_2\rho_i \ g \left(h - h_{NPL}\right)\frac{\left(T_i - T_o\right)}{T_o} \tag{8-2}$$

where
Δp_s = pressure difference due to stack effect (in., w.c.)
ρ = air density, lb_m/ft^3 (about 0.075)
g = gravitational constant, 32.2 ft/s^2
h = height of observation, ft
h_{NPL} = height of neutral pressure level, ft
T = absolute temperature, °R (= °F + 460°)
C_2 = unit conversion factor = 0.00598
Subscripts
i = inside
o = outside

This equation can be simplified to yield an estimate of the magnitude of the stack effect on a building, in inches of water column, equal to:

$$\Delta p_s = 1.44 \times 10^{-2}\left(h - h_{NPL}\right)\frac{\left(T_i - T_o\right)}{T_o} \tag{8-3}$$

This estimate neglects any resistance to airflow within the structure. However, since tall buildings have elevator shafts, this is a reasonable assumption to apply. There is also the uncertainty of where to place the neutral pressure level (NPL). It is reported[2,3] that the NPL in tall buildings varies from 0.3 to 0.7 of the total building height, although another source[4] pegs it at 0.7.

Example 8.2. For a building 546 ft tall, based on 42 stories each having a height of 13 ft, an indoor/outdoor temperature difference of 50°R, and an assumed NPL at the mid-height of the building, the resulting calculation can be performed:

$$\Delta p_s = 1.44 \times 10^{-2}(546 - 273)\frac{(530 - 480)}{480} \tag{8-4}$$

$$\Delta p_s = 1.44 \times 10^{-2} (273) \frac{50}{480} \qquad (8\text{-}5)$$

$$\Delta p_s = 0.410 \text{ in., w.c.} \qquad (8\text{-}6)$$

0.41 in. w.c. is equal to 2.13 lb/ft^2.

Another published equation for computing the magnitude of the pressure difference in inches of water column (in., w.c.) due to this stack effect uses the following equation:[4]

$$\Delta p_s = 0.84 \frac{(\beta h)^{1+n}}{1+n} \left(\frac{1}{T_o} - \frac{1}{T_i} \right) \qquad (8\text{-}7)$$

where
β = ratio of height of NPL and building height = 0.7
n = flow exponent = 0.65 for all classes of construction

Applying the same input variables with this equation as used in Example 8.2 leads to the following result:

$$\Delta p_s = 1.30 \text{ in., w.c.} \qquad (8\text{-}8)$$

The discrepancy between these two results indicates that the calculation of the magnitude of the thermal stack effect is not an exact science. Whatever the actual value is, know it to be significant in heated buildings. One reason for the uncertainty as to what the actual magnitude of the stack effect is that it will vary as a function of the amount of leakage occurring. The leakier the upper portion of a building is, the greater the need for replacement air at the bottom of the building and consequently the greater the suction (i.e., pressure differential) at that location.

An example of this situation exists in radon mitigation efforts. One of the indicated steps as part of an effort to reduce elevated radon levels in residential structures is to reduce the upper level "thermal bypasses" (i.e., leakage sites) in order to reduce the negative pressures occurring in the basement. It is, after all, these negative pressures that are the driving forces for sucking the radon-containing soil gases out of the ground and into the building.

This stack effect has been identified as being responsible for the transport of air contaminants in many building investigations.

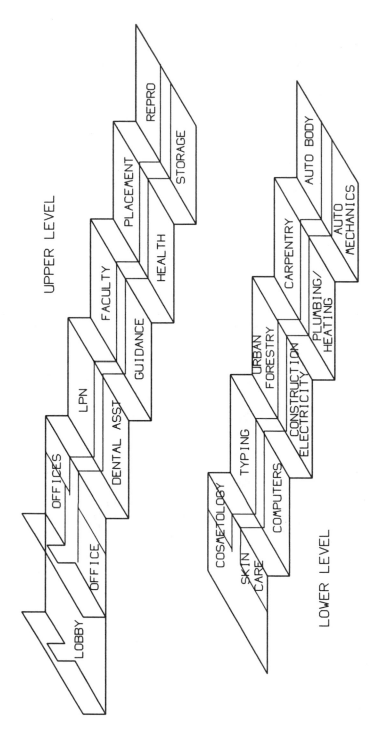

Figure 8.1   Floor plan of vocational high school on side of hill.

Example 8.3. In one building, with perimeter heating but without a central HVAC system, complaints that odors from the dry cleaners on the west side of the first floor of the building were being detected on the east side of the second floor. The confusion among the workers was due to the fact that the dry cleaners were on the west side of the building, but the odors were noticeable on the east side of the second floor, especially in and around the bathrooms. With ladder and air current tubes, the air movement pattern through the building was identified as traveling across the first floor ceiling, parallel to the floor joists, traveling from above the dry cleaning operation easterly to the plumbing penetrations in the floor of the second floor bathroom. Once the pathway was identified and understood, the most effective location for eliminating this pathway was identified and the situation mitigated. In this case, it was the penetrations in the floor around the pipes going to the bathroom that were sealed off to eliminate this pathway. The underlying cause for this pathway was, of course, the stack effect operating on the building; this could not be eliminated. Although there was lateral movement across one floor of the building, this air pathway was still tied in with the thermal stack effect, which was creating an upward flow of air through the building any way that it could.

Example 8.4. This example also involved a heated building that did not have a central HVAC system. This investigation involved a high school level vocational school building that was built on the side of a steep hill. A floor plan for the building is presented in Figure 8.1. This building had a central corridor that

Figure 8.2 Dirt deposit patterns on ceiling tile.

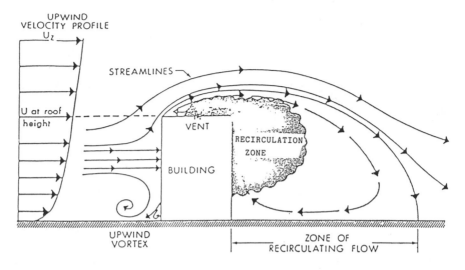

Figure 8.3   2-Dimensional recirculation pattern around building.

functioned as a conduit for the transport of air contaminants generated from activities occurring at lower levels in the building up to upper-level office spaces. The specific pathway between the "dirty" activities (i.e., the auto mechanics shop) and the central corridor involved the fact that the openings which existed where the corrugated roof decking rested on top of steel beams that ran the length of the building had not been sealed. Although the central corridor had a dropped, suspended ceiling, it was open above to the roof decking, as were the activity areas along both sides of the corridor. The identification of this specific pathway was achieved by visual observations, aided by the use of an air current. The clincher in this determination was the presence of dirt residues deposited along the edges of the ceiling tiles above the central corridor. As shown in Figure 8.2, the dirt was deposited on these tiles where the contaminated air had been drawn down through this suspended ceiling.

REENTRAINMENT OF EXHAUSTS FROM STACKS

In addition to pressures across the building shell created by temperature differences, there is also the action of the wind that can have an influence on the transport of air contaminants. The importance of this influence is not in the pressure differentials created across the building shell, but in the zones of recirculation created by the interaction of the building itself and the wind. These zones of recirculation, as part of the aerodynamic wake of the building, can trap the contaminants from building exhausts and permit them to reenter the building via either unintentional or intentional air intakes into the building or the HVAC system. Figures 8.3 and 8.4 present sketches of recirculation patterns created by

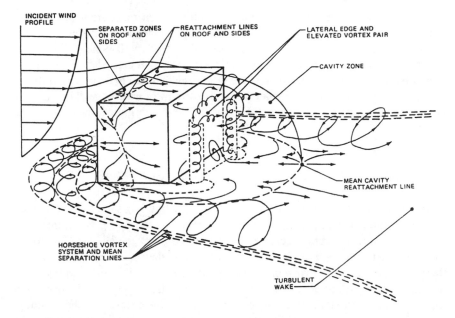

Figure 8.4   3-Dimensional recirculation pattern around building.

the interaction of the wind and a building. Equations for estimating the dimensions of the rooftop recirculation cavity have been worked out by Wilson and Britter.[5] These dimensions vary roughly in proportion to the square root of the area of the upwind face of the building. In evaluating the potential location of air intakes for a building to be built, it is therefore recommended that IAQ considerations be included in this aspect of the building's design by making sure that neither the air intakes nor building exhausts are located within the boundaries of these recirculation cavities. The reentrainment of air contaminants from exhaust plumes back into the building, after all, represents a pathway of potential odor transmission. In situations where this type of problem exists, other equations by Wilson[6] can be used to compare the various options for reducing the reentrainment potential. These options include:

1. reducing the source strength by process modification
2. increasing the stack height
3. increasing the stack discharge velocity
4. relocating the stack
5. relocating the air intake
6. prediluting the source with outdoor air prior to its release into the atmosphere

In addition to air contaminants reentering the building at the outdoor air (OA) intakes of the air handling units (AHUs), another route of reentry is through unintentional air intakes that may exist. Unintentional air intakes can exist whereever there are penetrations in the building envelope that are at a negative pressure with respect to the outdoors. This unintentional introduction of outdoor air into the building at lower levels can be aggravated by the location selected for the mechanical room. This is because the mechanical room will typically be under negative pressure with respect to the outdoors due to unsealed metal-to-metal connections and penetrations in the AHUs. This also means that air contaminants arising from sources located in the mechanical room or nearby, such as loading docks, can be drawn into the HVAC system and distributed throughout the building.

HVAC EQUIPMENT-RELATED PRESSURE RELATIONSHIPS

In addition to the pressure differentials across the building shell that are created by indoor/outdoor temperature differences and the action of the wind, there are also the pressure differentials created by the operation of the HVAC equipment. The operation of this equipment can either aggravate the conditions created by the thermal stack effect or it can work to reduce their effects. The major indicator of how the HVAC system or systems are functioning in this respect is whether or not the occupied spaces of the building, especially at the lower levels of the building, are being maintained at a positive pressure, typically 0.05 in. water column (0.26 lb/ft^2, 12.5 Pa, or 0.0018 lb/in.^{2}), with respect to the outdoors. The achievement of this positive pressurization of the occupied spaces with respect to the outdoors is important with respect to the quality of outdoor air being delivered to the building occupants. In a building with a properly functioning HVAC system, the maintenance of the occupied areas of the building at a positive pressure prevents or minimizes the infiltration of unconditioned air into the occupied areas of the building. That is, if outdoor air is permitted to infiltrate into the building at the perimeter, it will not have the benefit of thermal conditioning or filtration and can therefore lead to drafts in the winter or the excessive transport of dirt from the outdoors. It can also be a source of odors, depending on the activities occurring in the vicinity of the building. Another result of this pressure imbalance is that it can affect the distribution of outdoor air in the building.

Example 8.5. In one building investigation, it was determined that the occupied areas of the building were operating at a negative pressure with respect to the outdoors. The surrounding environment was relatively pristine so there was no noticeable problem with the deposition of dirt or dust. The building shell was sufficiently tight, as determined from CO_2 data that was collected during the unoccupied hours from 6:00 p.m. to 11:00 p.m. with the HVAC equipment turned off. This CO_2 data, which yielded a straight line when plotted on a log-linear graph, indicated a natural ventilation rate of 0.068 air change per hour (ACH) on January 3, 1991, 0.065 ACH on January 4, and 0.10 ACH on January 7. The OA temperatures for these evenings were in the low 30s on the 3rd and 4th, while on the 7th the temperatures dropped down to the 20s, resulting in the increased ventilation rate due to increased driving force for infiltration and exfiltration.

It was determined that although enough outdoor air was coming into the building at the perimeter, and providing ventilation there, interior areas that were only served by the HVAC equipment were not getting enough outdoor air for ventilation. This was determined by measured CO_2 concentrations in excess of 1000 ppm in interior occupied conference rooms. CO_2 concentrations measured in the perimeter areas, however, remained below 875 ppm. Outdoor levels were 350 ppm.

The pressurization of the occupied spaces with respect to the outdoors is typically achieved by operating the supply fan such that it attempts to introduce a larger volume of air into the building than the return air (RA) fans and exhaust fans are attempting to remove from the building. Since the quantity of air that can be introduced into the building must equal the quantity leaving, a net overpressure is achieved in the occupied areas of the building.

Because of this goal of achieving a positive pressure of the occupied spaces with respect to the outdoors, one of the first steps in the evaluation process is therefore the measurement of the pressure relationships that exist across the walls of the building, between the indoors and the outdoors. When documenting the pressure relationships that exist in a building, it should be noted that other portions of the building envelope (such as where the return plenum is adjacent to the outdoors) will therefore be under a negative pressure with respect to the outdoors, and therefore have the potential for the unintentional introduction of outdoor air. Since this outdoor air would be entering the return plenum, it would then travel back to the AHU, where it would be thermally conditioned and filtered by that system; however, it still could be a source of odors or insects.

Example 8.6. In one building under evaluation, where the return plenum portion of the building extended out over the occupied space below, gaps between the architectural building panels permitted not only the unintentional introduction of outdoor air into the return plenum, but also insects. These insects then pro-ceeded to accumulate and die in the nearest perimeter lighting fixtures. It was the presence of these dead insects in the light fixtures which was one of the complaints that was mentioned in the initial briefing to document the concerns of the building occupants. Due to the lack of accessibility to the building's overhang from the outside, the investigation into the performance of this building required climbing up into the return plenum leading to the identification of the gaps in the panels. Here, then, although the pressurization of the building was as it should be, the lack of attention to detail in the installation of the exterior architectural panels led to concerns of inadequate IAQ in the minds of the building's occupants.

In addition to the pressure differential across the building envelope, another important pressure relationship is the one that exists in the mixing box of the

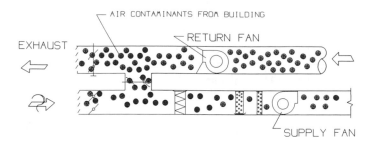

Figure 8.5   Outdoor air intake functioning as building exhaust.

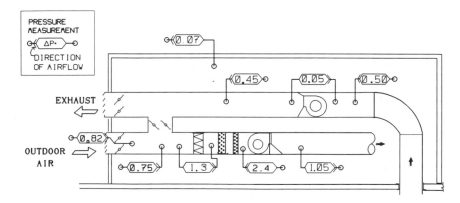

Figure 8.6   Pressure measurements at AHU with OA dampers closed.

AHUs. This pressure relationship is significant because it can affect the ultimate delivery of an adequate quantity of outdoor air to the building occupants. Instead of being positively pressurized, this location needs to be negatively pressurized in order to draw the outdoor air past the OA dampers and into the AHU. In some buildings, the selection of the rotational speeds of the RA and supply air (SA) fan failed to consider the buoyancy of the RA stream, a function of the building's *stack effect*. This fan imbalance can then lead to a pressure imbalance in the building, where the mixing box is positive with respect to the outdoors. In this situation, the OA intake ends up merely as a building exhaust. This condition is depicted in Figure 8.5. A less serious imbalance can also exist where the intended building exhaust location is in fact functioning as an unintentional OA intake.

Therefore, as part of the documentation of the performance characteristics of HVAC systems, there are several pressure relationships at the AHU that should be measured. Understanding these pressure relationships helps predict the performance characteristics of the HVAC equipment. The pressure relationships that can be measured for typical AHUs are presented in Figure 8.6 with a set of actual measurements. These measurements were obtained in the early evening with the fans still running, but with the OA dampers in their "closed" position. This building was being operated with the fans on for most of the night, with some of

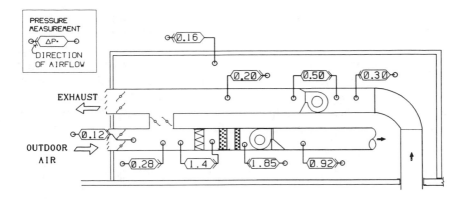

Figure 8.7   Pressure measurements at AHU with OA dampers at a minimum.

the units cycling off periodically, but with the intent that there would not be the introduction of outdoor air at this time. In this figure, there are several interrelated pressure differences. For instance, the 0.82 in. w.c, the pressure drop across the OA intake, minus the 0.07 in. w.c. pressurization of the mechanical room to the outdoors, is equal to the 0.75 in. w.c., the pressure between the mechanical room and the mixing chamber of the AHU. As mentioned elsewhere, the amount of air passing through the OA dampers is a function of both their net open area and the pressure drop across this opening. Even with these dampers in their "closed" position, however, they were sufficiently leaky and the pressure drop was sufficiently high that the amount of outdoor air entering here represented 20% of the supply air. Admittedly, since this 20% determination was based on mixed air, return air, and outdoor temperature apportionment, there is some uncertainty in the actual measurement; it can, however, be stated that the intent to exclude outdoor air under these conditions was not being achieved.

Another significant relationship is the measurement of the pressure drop between the mechanical room and the outdoors. This is significant in that most mechanical rooms I have observed are at a negative pressure with respect to the outdoors. There are two reasons why this one is positively pressurized. One reason is that it is a penthouse location and thus the stack effect is working to pressurize it. This differs significantly from a basement location where the stack effect is working to depressurize it. The second reason is that one of the AHUs has an open grille between its positively pressurized return air ductwork and the mechanical room. The introduction of this air into the mechanical room was able to exceed the leakage into the AHUs.

Another group of interrelated pressure measurements are those three around the return air fan. Here the 0.05 in. w.c. overpressure across the return air fan when added to the 0.45 in. w.c. suction downstream of this fan equals the 0.5 in. w.c. suction upstream of this fan.

For the same AHU as displayed in Figure 8.6, Figure 8.7 displays the pressure relationships with this system adjusted to their minimum outdoor air position. Although the pressure drop across the OA intake dampers has dropped down to

0.12 in. w.c., the increase in the opening of these dampers has increased enough so that the measured OA percentage was estimated at 39%. This change in damper position has also resulted in changes to most of the other measurements as well. While two examples of pressure relationships from just one AHU do not really begin to cover the complexity of these relationships, these examples do point out the amount of potential variation that can occur and the need to document these pressure relationships and the position of the system dampers as part of an investigation of the performance characteristics of HVAC systems.

ASSESSMENT TECHNIQUES FOR IDENTIFYING AIR MOVEMENT PATHWAYS

As already mentioned, the identification of air movement pathways in a building begins with a visual assessment of the evidence, added to by the use of air current tubes. One useful procedure is to have floor plans reduced to clipboard size and go to each doorway in the zone of concern, release a puff from the air current tube at that location, and then enter an arrow on the floor plan indicating the direction of air motion. At each location, this arrow will not only indicate the local direction of air movement but, because the air will always be moving from a zone of higher pressure to a zone of lower pressure, the accumulated arrows will also indicate the basic pressure relationships that exist at the zone of concern. This assessment technique will determine if a given zone is positively or negatively pressurized with respect to the areas surrounding it. This is just a qualitative assessment. In order to quantify the magnitude of this pressure differential, some sort of pressure measuring device will need to be used, such as an aneroid gauge like the Maghehelic™ gauge or a digital micromanometer.

Recently Renovated Office Area

Example 8.7. In one investigation, the complaints in the building were limited to a group of people working in an interior portion of the building. One of the initial steps was therefore to go to each of the doorways at the perimeter of this zone to determine if airflow at these locations was into or out from this zone to the surrounding corridor. The results of this determination are presented in Figure 8.8. The resulting outward flow of air at each of the doorways between this interior zone and the surrounding corridor indicated that this zone was positively pressurized with respect to the corridor. The significance of these results, the airflow being out from the zone to the corridor, is twofold. First, it indicates that the HVAC system serving this area is pressurizing it with respect to the surrounding area; and secondly, the search for sources of air contaminants should therefore focus on the HVAC system and the space itself, with reduced initial concern as to the other activities occurring on this floor of the building.

The HVAC system was in fact a source of air contaminants in this situation. During the recent renovation of this area, the ductwork had not been sealed off

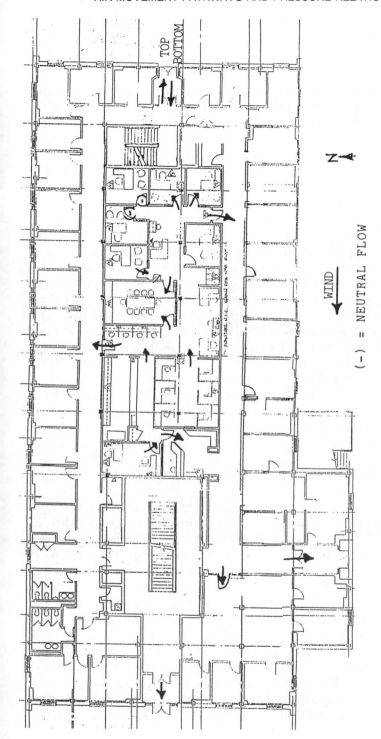

Figure 8.8   Floor plan showing air movement patterns: Example 8.8.

from this activity and the resulting debris, such as gypsum dust, had been able to accumulate in the ductwork. This in itself did not need to be a problem if the ductwork had been cleaned prior to occupancy, but this step had not been taken either.

If the identification of air movement patterns and an understanding of the basic pressure relationships are not sufficient to establish the specific location of the pathway of odor transmission, the next step that can be taken is the performance of tracer testing. Sometimes, for instance, the pathway is not accessible to direct observation as it was in the above cases. For instance, it may travel within walls or involve larger distances. This is where the use of tracer testing can be very useful.

In this approach, after background checks have confirmed that there is no detectable tracer response, a tracer release is initiated at or near the location of the suspected source of air contaminants. This background check is very important because, in one hospital odor investigation, it was determined that CAT scan equipment in the basement of the building being tested was using sulfur hexafluoride (SF_6) as a high-voltage RF shield. Since this installation was not leak-tight, SF_6 was present at low levels at various locations in the building. After the tracer release has been initiated, sampling begins at the location where the complaints are occurring and at locations between this location and that of the suspected source. Analysis of the tracer concentration in these samples is then performed. The conclusions are derived from the fact that the sampling locations where the tracer shows up most rapidly and in greatest concentrations are the locations most intimately connected with the air movement pathway of interest. Tracer testing has the additional advantage of not only be able to identify the location of specific pathways of air movement, but it can also permit the quantification of airflows and dilution amounts as well.

Odors from Sterilizer Serving Animal Farm in Hospital

Example 8.8. In one building, the time of the complaints was checked with other activities at the particular hospital and found to correlate directly with when the sterilizer serving a colony of immune-suppressed, naked mice had just finished sterilizing a batch of their food. This activity was a source of very strong odors that were detected in several locations in the hospital. Based on the record of complaints, this relationship did not vary with the wind direction, indicating that the pathway of odor transmission was more direct than would be expected for just the reentrainment of exhaust air via the outdoors. In discussions with the person responsible for the operation of this "animal farm," it was determined that because of the need to maintain a sterile environment for the mice, the "clean" side of the sterilizer was ventilated by a separate system to the one which served the "dirty" side of the sterilizer. The testing protocol therefore involved separate tracer releases at each of these two locations. Some of the results from this testing effort are presented in Figure 8.9. The mouse colony and its sterilizer are located on the

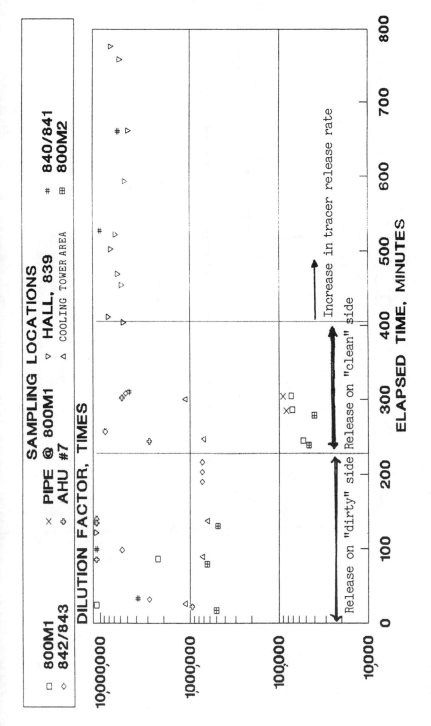

Figure 8.9   Tracer measurements from mouse colony odor study.

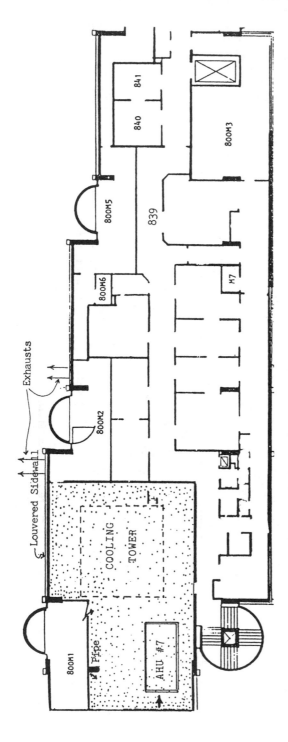

Figure 8.10 8th-floor plan for level above mouse colony.

seventh floor of this building. A portion of the eighth floor plan for this building is presented in Figure 8.10. The complaints had been registered on the second, fifth, sixth, and eighth floors of this building. However, the most frequent complaints were from an occupant in the room numbered 840/841. This complaint location is served by AHU #7, which only serves the 8th floor. The remaining six AHUs serving the rest of the building are all located in the mechanical room on the fourth floor.

In the plot of results presented in Figure 8.9, the X-axis is elapsed time (in minutes) since the first release of tracer, while the Y-axis represents the log of a "dilution factor" of the tracer measurements. This "dilution factor" is derived from a calculation of the amount of air that would be required to dilute the amount of pure SF_6 tracer released down to the concentration amount detected in the air samples. The advantage of summarizing the results in terms of this dilution factor is that this approach permits the direct comparison of tracer measurements obtained during varying tracer release amounts, since the release of tracer was increased after 405 minutes into the test.

Another change in the tracer release occurred at 230 min into the test, when the tracer release location was changed from the "dirty" side of the sterilizer to the "clean" side of the sterilizer. As can be observed in Figure 8.9, this change resulted in decreases in the dilution factor between these releases and the samples collected in the eighth floor mechanical rooms. Remember, decreases in the dilution factor correspond to increases in the measured tracer concentrations, and a more direct relation to the tracer source. In addition to the tracer results, it was also determined that the mechanical room designated 800M1 was positively pressurized with respect to the outdoors. This inspection also identified the existence of penetrations in the walls of Room 800M1 which were leaking air into the cooling tower area. This cooling area, displayed in Figure 8.10 with the speckling, is surrounded by either vertical louvers or solid walls. A review of the tracer data, at 300 min into the tests, documents the pathway involved: from the sterilizer, into the mouse colony, into mechanical room 800M1, into the cooling tower area, into AHU #7, and then distributed to the rooms on the eighth floor including the complaint location.

Having identified the specifics of this pathway of odor transmission, the necessary requirements for eliminating this pathway become clear. The first recommendation is to seal up the penetrations in the walls of mechanical space 800M1. The second recommendation is to install an exhaust fan in the roof of Room 800M1 to make this space neutral with respect to the outdoors. The third step was to relocate the OA intake of AHU #7 sideways beyond the louvered enclosure of the cooling tower bay. This recommendation was included because of the possibility that the some of the exhaust from the mouse colony could be reentrained into the cooling tower enclosure during certain wind regimes, and also because IAQ considerations stipulate that no air intakes be in the vicinity of cooling towers. The reasons for this are that the water droplet drift from cooling towers can dampen filters in the AHU which would permit them to become sources of microbiological growth, or the drift itself could be contaminated with microbes. The configuration of the air intake should involve not only that added ductwork be outboard of the louvered wall, but it should also be extended downward to gain additional vertical distance away from exhausts, which would help to minimize the potential for reentrainment.

In terms of the pathways to the other locations in the building, the presence of tracer was also detected in the fourth floor mechanical room. The pathway between the eighth floor and fourth floor mechanical spaces was via the vertical mechanical shafts and the driving force was the pressure gradient between these two locations; the eighth floor mechanical space was more positively pressurized than the fourth floor mechanical room. In the fourth floor mechanical room, penetrations from the mechanical room in the AHUs, on the suction side of the fans, were also identified. This completed the identification of the existence of pathways to the other locations in the building. Efforts to eliminate this pathway involve exhausting air out of the tops of the vertical mechanical shafts to prevent the downward migration of odors and the more rigorous sealing of exhaust fans located in mechanical space 800M2.

Odors from Loading Dock at Health Care Facility

Example 8.9. One variation on the continuous release of tracer to document the existence of pathways of odor transmission is the use of a pulse injection of tracer. In my first tracer gas study to determine possible pathways of transport of air contaminants from a loading dock, which I performed in April 1981, the release was achieved by emptying the contents of a 20-cc syringe filled with pure SF_6. As I refined the technique over the years, my tracer release methodology now includes a tank of pure SF_6, connected to a pressure regulator, a needle valve, a calibrated flow meter, and some distribution tubing. This current arrangement provides more flexibility, but the original technique still got the job done. In this investigation, the occupants of one office area in a health care facility were complaining that air contaminants from the loading dock, which was outboard and below their location, were reaching their space. After these women had complained for some period of time, the height of the air intake to the AHU serving this office area had been increased. Despite this expenditure of money, to everyone's disappointment, the problem persisted. I was then hired to investigate other possible pathways of transport of air contaminants from a loading dock to this location. The first step in the investigation was the use of air current tubes to see where air might be entering the building at the loading dock. This inspection identified several penetrations in the ceiling above the loading dock which were drawing from the loading dock into the negatively pressurized mechanical room above. Fourteen tracer releases were performed on 2 separate days. Prior to the release of any tracer, however, samples were first taken to confirm the absence of any tracer measurements above the minimum detection limit of the test equipment. Release #1 in the mechanical room, followed by detection of tracer in the

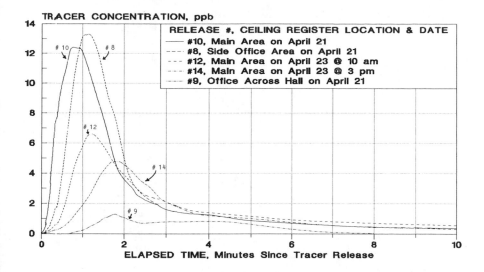

Figure 8.11 Tracer measurements from releases at loading dock.

supply air and near the exterior wall, confirmed that a pathway did in fact exist between the mechanical room and the complaint location. Test #2, which involved sampling at numerous locations in the office space, indicated that the SF_6 coming through the wall could be detected most strongly at the baseboard heaters. Test #3 was a negative determination and indicated that a particular open duct in the loading dock area was not involved in the pathway. Test #4 determined that the concentration of SF_6 passing through the wall peaked at about 11 to 20 min after the tracer release. Test #5 indicated that the concentration of SF_6 measured in the ceiling supply air register peaked in 2 min or less. Test #6 was another negative determination and determined that the basement area inboard of the loading dock was not involved in the pathway. Test #7 involved the release of tracer at a pipe penetration in the ceiling of the loading dock and, since tracer was again detected at the complaint location, this determined that this location was also part of the pathway of concern. This release location was repeated for tests #8, #9, and #10. The sampling for test #8 was at the supply register in the complaint area closest to the AHU. This complaint area was the first location served by this AHU. This concentration peaked at 13.2 ppb, 1.1 min after the tracer release. The sampling for test #9 was at the third supply register away from the AHU and was in a separate non-complaint office across the hall. This result, also plotted in Figure 8.11, indicated that the tracer was less than 10% of its concentration at the first supply register. The result for test #10, with its sampling at the second supply register, which was also in the complaint area, basically duplicated that for the first supply register.

These specific results are typical of a situation I have also observed elsewhere; that is, air contaminants entering an AHU just before the fan do not have the opportunity to become well mixed as they transit the supply fan. This pocket of

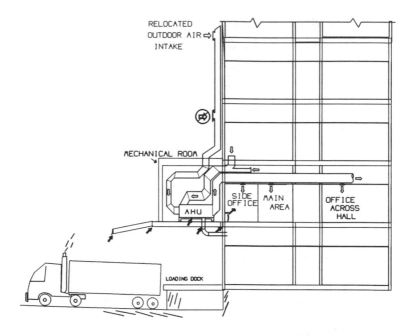

Figure 8.12 Building section showing pathways from loading dock.

contaminated air seems to be compartmentalized as it passes the fan and can emerge in a high concentration at just one or a few locations downstream, while they are not detected at other downstream locations.

Test #11 was a comparison test and it determined that tracer gas introduced below the floor of the mechanical room penetrated though the wall cavity more quickly than tracer gas introduced into the mechanical room itself. This test concluded the effort for the first day of on-site evaluations. This testing had documented that there were two mechanisms by which fumes from the loading dock could be transported to the office space where the complaints were occurring. These pathways are indicated in Figure 8.12. The pathways from the loading dock to these offices begins with the penetrations in the ceiling above the loading dock. This load-bearing platform, which constitutes the mechanical room floor and the loading dock ceiling, contains hollow spaces and voids which permit the air drawn in from below to enter both the wall cavity of the main building and the mechanical room itself. From within the wall cavity, the penetrations associated with the piping to the baseboard heaters complete this path to the office space. The driving force for this pathway into the space is the building's stack effect. The completion of the pathway into the mechanical room involves the entry into the conditioned air stream via unsealed metal-to-metal joints at the bottom of this AHU. The driving force for this pathway is the suction created by the supply fan.

Although there are not many practical options for eliminating these driving forces in this situation, efforts to reduce the magnitude of the pathways by the sealing of penetrations were begun immediately after this day of testing. After 1 day's effort of sealing penetrations, testing was performed to evaluate the effectiveness of these efforts. This evaluation effort involved the repeat of the procedures involved in test #10. The results from the morning test, test #12, are presented in Figure 8.11. These results determined that the initial sealing efforts had restricted the pathway in that there was less tracer reaching the office space. The interval from 10:30 a.m. to 2:30 p.m. was then devoted to sealing up leakage sites that were still detectable with the aid of the air current tubes. The effectiveness of this additional sealing effort was demonstrated by the afternoon tracer tests, tests #13 and #14, where the peak height was further reduced. The increase in the time to maximum peak height (see Figure 8.11) also indicated that the pathway had now become more arduous for the transport of air from the loading dock.

This example points out some of the challenges of eliminating IAQ problems that result from a failure to fully incorporate IAQ considerations into the building design process. At issue here is the decision to use hollow precast concrete planks to support a mechanical room located directly above a loading dock. There is also the installation issue of the failure to seal the bottom metal-to-metal joints of the AHU at the time of construction because of the difficulty of achieving an adequate seal at this later date. Because of the potential for this particular problem to reoccur in the future and because the driving forces will continue to exist and the effectiveness of the seals can be expected to deteriorate somewhat, there are other mitigation options that can be considered. One approach would be rigorous enforcement of a regulation prohibiting the idling of vehicles at the loading dock. Another solution would be to install a system of flexible rubber hoses, an exhaust fan, and a stack which would collect these vehicle emissions at their source and discharge them away from the loading dock and above the mechanical room. These test procedures also provide an example of the trial-and-error steps required for evaluating hypotheses relating to IAQ.

Evaluation of Pathways from Loading Dock to Office

Example 8.10. Eight years later, in July 1989, I was hired to perform yet another building investigation involving concerns about pathways of air movement from a loading dock. This situation is in some way similar yet different from the previous example. It is also interesting to note how the tracer test procedure is improved by the use of the continuous release approach. The background to my involvement in this project was that after a history of problems relating to the presence of odors from the loading dock, the primary focus of the study was an area that had just been renovated and had improvements made to its HVAC system to prevent future problems. I was hired to assess the effectiveness of these improvements. Since the history of problems included the infiltration of odors directly into the space, the first step in the study was the assessment of air movement patterns in this zone to determine if it was now positively pressurized

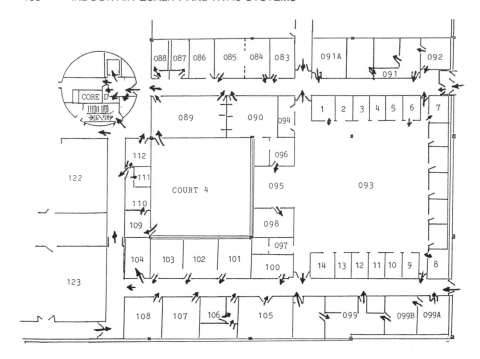

Figure 8.13 Floor plan showing air movement patterns: Example 8.10.

with respect to its surroundings. The results of the air movement assessment is presented in Figure 8.13. As can be observed from this figure, the office spaces are positively pressurized with respect to the corridors, but the corridors are negatively pressurized with respect to the outdoors. This being a 22-story building, the building's negative pressurization on the second floor can of course be attributed to the thermal stack effect, as indicated by the flow into and up the circular utility riser shown in Figure 8.13 as "CORE D." Even though the office spaces were positively pressurized with respect to the building corridors, locations 109 and 089 were negatively pressurized with respect to the outdoors, as evidenced by the airflow direction made between them and the outdoors at Court 4 (see Figure 8.13).

The investigation into whether or not a pathway existed between these office spaces and the loading dock involved the release of SF_6 tracer gas at the loading dock, with subsequent sampling at various locations in and around the building. In locations where the presence of tracer is detected, the higher its concentration the closer it is to the actual pathway of air movement in the building. The release of tracer was achieved with a tank of compressed, pure SF_6 connected to a pressure regulator, a controlling needle valve, a calibrated flowmeter, and a length of tubing. This tracer release setup was located in a locked car parked in the loading dock area, with the release tubing extending out of the car via the sunroof. Sampling for the presence of tracer was achieved manually with Tedlar gas sampling bags and a squeeze bulb.

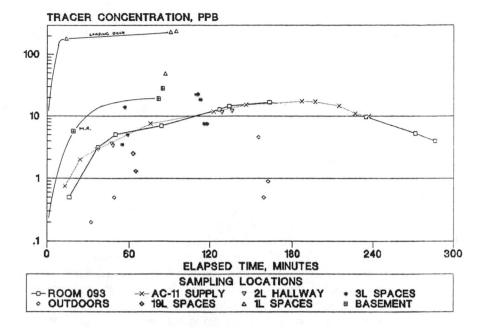

Figure 8.14 Tracer measurements from releases at loading dock.

The results of the tracer measurements are summarized in Figure 8.14. After confirming the presence of no detectable background concentrations, the tracer release rate initially started at 40 cc/min at 11:02 a.m.; 25 min later, a measurement of 1.0 ppb was detected in the supply air to Room 093, documenting the existence of a pathway of air movement from the loading dock to the office area of concern. The tracer release rate was then increased to 220 cc/min at 11:40 a.m. At 11:47 a.m., the measured tracer concentration at the loading dock, midway between the tracer source and a corridor opening into the building, was 182 ppb. This concentration increased slowly, up to 233 ppb at 1:04 p.m. and 242 ppb at 1:08 p.m. In the context of this analysis, relating the measured concentration to the release rate permits the calculation of a dilution volume. A volumetric flow rate of 1000 m³/min is required to dilute a pure tracer release of 220 cm³/min down to a concentration of 220 ppb. Therefore, prior to entering the door separating the loading dock from the interior of the building, the tracer released at the car was being diluted into a volumetric flow of approximately 35,310 cubic feet per minute (cfm). These loading dock measurements are located in the upper left of the tracer plot displayed in Figure 8.14. The next highest concentration displayed in this plot, also indicated with the Δ, is for the measurement of 49.4 ppb obtained next to the elevator door on the first floor, the same floor that the loading dock is on. The dilution of this tracer concentration, as compared with that at the loading dock door, can be interpreted to indicate that only slightly more than 20% (49.4/233) of the air being drawn to this elevator shaft is coming from the loading dock. The concentration next to the elevator shaft at the basement level, one floor below, was determined to be 28.2 ppb at 12:58 p.m. The reduction in tracer concentration from 49.4 ppb entering the elevator at the first floor down to 28.2 ppb leaving the elevator shaft at the basement can be explained by the observed introduction of additional dilution air in the elevator shaft on the second floor.

The build-up of tracer in the basement mechanical room (5.7 ppb at 11:52 a.m. and 19.3 ppb at 12:55 p.m.) was influenced by the estimated volume of 2.7 million ft^3 in this space. From the mechanical room, the room air was drawn into the AHUs at observable penetrations. Specific leakage sites included the filter access panels, fist-sized holes in the vibration isolation boots, uncapped sampling ports (for airflow rate testing), and at metal-to-metal joints. After the basement measurements, the next highest concentrations of tracer were observed in the library on the third floor, served by AC-15. After this, the next highest concentrations were found in the original office area of concern, served by AC-11. During the testing effort, the determination of higher concentrations in the area served by AC-15, as compared with those in the AC-11 zone, was accomplished prior to the inspection of the condition of these AHUs. Therefore, it was hypothesized that, upon inspection, there would be larger observable openings between the mechanical room and AC-15 than could be observed with AC-11. This hypothesis was in fact confirmed upon inspection: the hole in the AC-11 unit at the flexible vibration isolator was the size of my clenched fist; while at the AC-15 unit, the hole was twice that size.

Another comparison of relative tracer concentrations, between the mechanical room and the supply, permits a determination of the magnitude of the leakage occurring. It should be explained that this building had originally been designed with air intakes and fans in the penthouse and ductwork shafts down the core of the building to transfer this outdoor air down to the 28 AHUs in the basement mechanical room. However, due to unexpected problems of noise and vibration, the rooftop air intake fans were not being operated. To compensate for this, the supply fans in the basement mechanical room were sped up to be able to overcome the resistance of the air intake shafts by themselves.

A comparion of tracer concentration measurements in the mechanical room and supply air to Room 093, at approximately the same time, permits the estimation of the percentage of air going to this area that has originated in the mechanical room (M.R.). At 11:52 a.m. and 11:57 a.m., the comparison is between 5.7 ppb in the M.R. and 2.0 ppb in the supply air. This relationship leads to the conclusion that 35% of the supply air is coming from the M.R. A repeat of this comparison, at 12:55 and 12:49 p.m., has 19.3 ppb in the M.R. and 7.6 ppb in the supply air. This leads to the conclusion that 39% of the supply air is coming from the M.R. Each of these values is within 6% of the mean of 37%. This is well within the 10% range expected for this type of experimental data.

At 2:30 p.m., the tracer release in the loading dock was terminated. After that, the measured tracer concentrations in the building began to decrease. The rate of decrease in the tracer concentration in Room 093 corresponded to a ventilation rate of tracer-free air of 1.0 ACH. The ability to obtain this additional information is just a fringe benefit of the use of SF_6 tracer testing as part of IAQ investigations in non-industrial buildings.

This IAQ problem can be explained by deficiencies in terms of the design, installation, operation, and maintenance of the HVAC system that have all contributed to the creation of a pathway of air movement from the building's loading dock to the occupied spaces in the building.

This chapter's intent was to explain the relationship between pressure relationships and air movement patterns in a building, as well as to describe how these conditions can be evaluated with the aid of common sense, air current tubes, and tracer tests. The next chapter expands on this approach by discussing the use of other tools and techniques.

REFERENCES

1. *ASHRAE 1989 Fundamentals Handbook*. 1989. American Society of Heating, Refrigerating and Air Conditioning Engineers, Inc. Atlanta, GA. p. 23.4.
2. Tamura, G. T. and A. G. Wilson. 1966. "Pressure Differences for a Nine-Story Building as a Result of Chimney Effect and Ventilation System Operation." *ASHRAE Trans.*, 72(1):180.
3. Tamura, G. T. and A. G. Wilson. 1967. Pressure Differences for a Nine-Story Building as a Result of Chimney Effect and Ventilation System Operation." *ASHRAE Trans.*, 73(2):II.1.1.
4. Shaw, C. Y. 1989. "Wind and Temperature Induced Pressure Differentials and an Equivalent Pressure Difference Model for Predicting Air Infiltration in Schools." *ASHRAE Trans.*, 86(1):268–279. No. 2573.
5. Wilson, D. J. and R. E. Britter. 1982. "Estimates of Building Surface Concentrations from Nearby Point Sources." *Atmospheric Environment*, 16(11):2631–2646.
6. Wilson, D. J. 1983. "A Design Procedure for Estimating Air Intake Contaminant from Nearby Exhaust Vents." *ASHRAE Trans.*, 89:2769.

CHAPTER 9

Evaluation Tools and Techniques

OVERVIEW

The tools used to evaluate the performance of ventilation systems range from merely educating four of the five senses of the human body to the use of sophisticated and expensive equipment such as those required for tracer testing. The four basic senses to be educated or "calibrated" by training and experience are those of hearing, seeing, smelling, and touching. The specifics for each of these innate human abilities are discussed in the portion of this chapter devoted to the investigator as instrument. Beyond the abilities we have been born with, there are many tools that can be used to augment what we can sense. There are the simple tools, such as the air current tube (or as some people call it the smoke pencil) which permits the direction of air movement to be visualized. There are basic tools that should be in a tool box and there are the pressure-measuring devices which permit pressure relationships to be quantified. Then there are the more sophisticated and therefore potentially more powerful tools that permit the measurement of either the carbon dioxide (CO_2) in the air or for measuring a tracer in the air, such as sulfur hexafluoride (SF_6).

THE INVESTIGATOR AS INSTRUMENT

The information gained from the four basic senses can be summarized as follows:

- Hearing → Knowing what to listen for and what to believe
- Seeing → Knowing what to look for
- Touch → Knowing what to feel for
- Smell → Recognizing odors

171

Hearing: Listening to the Occupants and Building Operators

The occupants of a building represent a significant potential source of information for the building investigator. Therefore, time should be allocated as part of the evaluation to listen to these people who have already been spending lots of time in the building that you are attempting to become familiar with in a very short period of time. Everything these people have to say is a potential clue as to what is going on with the building and its systems. Without going to the trouble and expense of administering a questionnaire, the basic questions to be asked include the nature, timing, and location of complaints. For example, with respect to the timing of complaints, the questions should determine the following. When did the complaints begin? Do the complaints vary by time of day, or day of the week, or by time of the year? For instance, in a recent investigation of mold allergy complaints, the question was raised as to when the complaints started. Hearing that this complaint coincided with the start of the cooling season (i.e., when the air conditioning went on) indicated that the drainage from the condensate pan needed to be checked. This subject was touched upon previously, with additional examples, in the beginning of Chapter 5 as part of the evaluation process: the review of occupant's complaints or concerns.

While everything these people have to say is a potential clue, it must be stressed that people's perceptions may be misleading. There are several examples of this phenomena. In one situation, as part of a total environmental assessment in a building, occupants were asked to rate how noisy they thought their offices were. These people perceived their offices to be noisy because voices carried and conversations could be overheard. The reality was that this office was very quiet. Therefore, in the absence of background noise, the spoken word could be heard for some distance and the people lacked acoustical privacy. By adding "white noise" to the background, these people gained acoustical privacy and they no longer thought that the building was too noisy.

Another perception/reality issue is that people experiencing sick building syndrome frequently complain that "there is not enough oxygen in here." The reality is that if there is a ventilation rate of 15 cubic feet per minute (cfm) per person and an oxygen consumption rate of 0.763 ft^3/hour (0.36 liters/min) for an activity level of 1.2 met, the oxygen level in the room will be reduced from 21% outdoors to 20.9% indoors.[1] This decrease represents a change of only 0.48%. What is being experienced by the individuals, however, is the build-up of bioeffluents; that is, chemical compounds given off from living beings. For example, with an outdoor CO_2 concentration of 330 ppm, and a ventilation rate of 15 cfm, the indoor concentration could build-up threefold to over 1000 ppm if the people remained around long enough. This represents an increase of over 300%. For bioeffluents without a significant outdoor air component, the relative increase would be even greater.

For additional examples of situations where people's perceptions differ from reality conditions, check back to the beginning of Chapter 5.

In addition to hearing what the people have to say, the HVAC equipment itself can also provide auditory clues as well. Whistling supply registers indicate distribution problems. Either the system is unbalanced and too much air is attempting to pass through the register, or someone has attempted to shut off a register that isn't designed to be shut off and air is whistling by.

While I find it useful to filter the information received from the occupants of buildings, I have found a greater need to be very skeptical about some of the information obtained from people responsible for the operation of the building. These people may believe they know a building and its system are performing, but they too can be mistaken.

In one building investigation, which was part of a proactive study for a major Fortune 500 company, the majority of the people responsible for the buildings at one campus were mistaken as to the expected operational mode of the HVAC system. This overall study, which involved the evaluation of a random selection of their facilities, included meetings with the key building operations people. At one meeting, held before the actual building inspection, the consensus was that, based on the 55°F outdoor air temperature, the buildings would be operating in economizer mode. Upon inspection, however, one the buildings being evaluated was in fact operating in its minimum outdoor air mode. The reality was that the single individual responsible for the day-to-day operation of this building had overridden the basic control protocol for the HVAC system. He did this because the system had a design flaw which was creating a maintenance headache for him. The humidification for this system had atomizing spray nozzles located in the mixing chamber, upstream of the filter bank. Not only is this a bad idea because of the potential for wetting the filters and the resulting potential for microbiological growth, but in the northern climate, the spray nozzles were experiencing frequent freeze problems.

The challenge in conducting interviews with people who think they know how HVAC systems are being operated requires a tactful approach in attempting to distinguish between what they "know" to be true and what they "believe" to be true.

The involvement of computer controls for building management systems has just added to this confusion. In another building investigation, the intent of which was to assess the amount of leakage past closed outdoor air (OA) dampers, the building management system gave instructions to the 21 air handling units (AHUs) in the building to close their OA dampers. The computer then indicated that this condition had been achieved. Despite the 13°F outdoor air temperature, the mixing boxes for these units were quite warm; except for one unit, that is. Upon entering the mixing box of this AHU, it became immediately clear to me that the OA dampers were very much open. The explanation for this discrepancy was that there was a local switch at this unit that had its controls disconnected from the main computer system.

These examples point out the need to question all of the information received during a building investigation, and the need to verify for oneself exactly what is going on with the building and its systems.

Hearing can also be useful in the inspection and testing of the condition of the HVAC equipment. Air leaking at gaps in the AHU itself or at the distribution ductwork can produce audible whistling that lets the investigator know that the leaks exist.

Seeing: Knowing What to Look For

Of all the senses, the sense of sight is easily the most important for performing indoor air quality (IAQ) evaluations. The investigator needs to cultivate a "calibrated eyeball" so that he or she knows *where* to look and *what* to look for.

The movement of dirty air has a strong tendency to deposit some of its dirt as it passes through small openings. I first noticed this many years ago when I was identifying pathways of heat loss in residential structures. Light switches that permitted the entry of heated room air on its way into the wall cavity and then up to the attic via the holes around the wires invariably had deposits of accumulated dirt in their crevices. In a very similar fashion, leaks in ductwork will also leave tell-tale deposits of dirt as the high-velocity air slows down.

In the absence of these clues, the visual inspection can be aided with the use of air current tubes which permit the qualitative determination of air movement direction and velocity.

A discussion of the importance of knowing where to look and what to look for was presented and discussed as Example 8.4 in Chapter 8. Applying the four-component paradigm presented in Chapter 1 to this situation, the source and the affected persons were easily identified. The driving force transporting the contaminants from the source to the people was the stack effect of the heated air rising through the building. The challenge was to identify the specific pathway. The evidence that the pathway existed was visually displayed by the dirt deposited on the edges of the ceiling tiles in the lower portion of the central corridor (see Figure 8.2).

There are many other visual determinations that should be included as part of a basic inspection of HVAC systems. The following list provides a summary.

1. Starting at the OA intakes, one should look to see if they are in fact open; an air current tube can also be used to determine if outdoor air is actually being drawn into the system.
2. Looking out from the air intakes, the presence of sources of air contaminants should be checked for. Typical sources potentially include nearby cooling towers, sewer vents, kitchen exhausts, and rooftop rain puddles.
3. Within the AHU, the filter rack should be checked for the full complement of filters that do not have any gaps between or around them. The filters themselves, unless they were changed recently, should look dirty on their upstream side. They, however, should not be deformed in any way. The filters will only be as effective as their installation.

4. The presence of operating pressure gauges, measuring the pressure drop across the filter bank, provides a good initial indication of the level of maintenance being provided on this system. Neatly displayed logs of past filters changes, which include the dates, pressure drops, and individual responsible for these filter changes, are examples of even greater attention to detail.

5. The inspection continues to the condensate drain pan. This should be free of both debris and standing water. The ease of access to this location provides additional visually determined information on the frequency and adequacy of the maintenance being provided. The presence of a water trap draining to the outdoors can also be checked.

6. Looking at the supply fan itself will provide an indication of the overall cleanliness of the system.

7. Downstream of the supply fan, at the beginning of the distribution ductwork, the condition of the duct liner, if one is present, should also be checked. This location is one where mold growth will typically begin due to the high relative humidities, accumulated dirt, and suitable habitat on the surface of the duct liner. Also to be checked for is the erosion of the duct liner arising from the situation where the supply fan has been sped up past the original design specifications in order to attempt to deliver larger quantities of supply air (SA) and thereby provide additional cooling capacity in the building. If the duct liner shows signs of erosion, the ultimate destination of the chunks of this material, typically fiberglass and liner, should be checked. Is it clogging perimeter reheat coils that are inaccessible for cleaning, or is it showing up on desks in the occupied spaces?

8. In the occupied spaces, the visual inspection should include the locations of the supply registers and return grilles, and looking for dirt deposits as well as their relative locations. Obvious sources of air contaminants should also be checked for, such as smoking (although that can be smelled as well) and copying machines.

Just as one needs to be skeptical of the information obtained by listening to the people responsible for the operation of the building and its HVAC system, one also needs to be skeptical of the apparent setpoints of system controls; these are frequently out of calibration. In one building investigation that William Turner and I performed, the one mixed air sensor/controller for the four AHUs in the building was visually determined to be set a 60°F. That meant that is was supposed to be letting in enough outdoor air to cool the return air from 70°F down to the setpoint of 60°F. Numerous people had previously looked at the dial of this controller and believed its setting. The placement of a remote-reading thermometer in the mixing chamber where the sensor for this controller was, however, indicated that it was 10° out of calibration and was actually seeking 70°F; thus, the OA dampers for all of the AHUs stayed closed all of the time.

While it is typically impractical to undertake to calibrate all sensor/controllers in a building, it must be remembered that the performance of IAQ building evaluations is still both an art and a science. The art comes from gathering enough information to form hypotheses about how the HVAC system and the building are performing, in order to guide the rest of the investigation, which then consists of performing good science.

Touching: What to Check with Your Hands

Sometimes you need to get your hands dirty when performing an IAQ evaluation, although it is more likely it will be your clothes — when you are climbing around HVAC equipment, duct chases, and mechanical rooms. The usefulness of the sense of touch lets you know, for instance, if there is a significant pressure imbalance in a building when you can feel the resistance in opening or closing doors. Sometimes, you can even feel the wind rushing through cracks around doors and windows, although the use of air current tubes will more consistently yield useful information in this respect.

The sense of touch can be used to check on the firmness of the connections on the linkages of the OA dampers. In some buildings, it was the looseness of these critical components that prevented the OA dampers from opening as much as they needed to be.

The performance of IAQ evaluations requires a definite "hands-on" approach: opening access doors into large AHUs, removing access panels in small AHUs, removing ceiling tiles, and taking measurements. A word of caution about safety needs to included. Reaching into HVAC equipment can be very dangerous. Power is transmitted from motors to fans by fan belts that can grab loose clothing and severely damage body parts. HVAC controls are frequently located in close proximity to high voltage electrical conections.

In addition to getting one's hands dirty, one should also expect to get tired feet. Evaluating a three-story building with 250,000 ft^2 of area and 19 AHUs can require that several miles be walked in one day.

Smelling: Recognizing Odors

The sense of smell can also be useful in IAQ evaluations. One technique for establishing the existence of pathways of air transport in a building is by releasing a distinctive source of odor, such as peppermint oil, at one location and checking to see where else this distinctive smell can be detected. Some investigators have also reported being able to get clues for characterizing IAQ problems from the persistence of perfume odors and the like. I can readily detect the tell-tale evidence of cigarette smoke, even in areas intended to be smoke-free. Sometimes, this is due to recirculation from other areas served by the same air handling equipment that is not restricted in this activity; and sometimes, this is due to people smoking just outside doorways and the odors are drawn into the building due to the stack effect.

IAQ TESTING EQUIPMENT

Various equipment is available to aid the investigator in the performance of IAQ evaluations and inspections. This sections discusses equipment according to the following list:

1. Air current tubes
2. Pressure-sensing devices
3. Flow measurement devices
4. Other useful tool box residents

Air Current Tubes

Air current tubes, or "smoke sticks" or "smoke pencils" as some people may call them, function by creating a visual plume in the air which then permits the direction and relative velocity of any air movement to be visually detected. One such device, the Dräger Air Current Kit, uses glass tubes filled with fuming sulfuric acid. They work by pulsing room air through the tube to create a visible acid vapor cloud from the reaction of the moisture in the air with the acid. Appropriate caution should therefore be employed when using this otherwise very useful tool because the contents are highly corrosive. Another similar device uses titanium tetrachloride, the visible cloud from which contains hydrochloric acid. Again, an appropriate amount of caution is indicated to keep the fumes away from people's breathing zones, especially your own. Although these devices involve small releases of acid, I still consider them more appropriate to use than a burning punk or cigarette. Not only does the smoke from a cigarette contain 4,000 compounds, several of which are carcinogenic, but also its smoke is heated, is thermally buoyant, and will tend to create its own rising air currents. The air current tubes, which rely largely on the moisture already in thermal equilibrium with the space, will be thermally neutral and can show evidence of downward motion where it exists.

In one investigation involving complaints of multiple chemical sensitivity (MCS), a colleague of mine, Terry Brennan, was understandably reluctant to use either of these devices and has reported getting useful information using the fluff from milkweed seeds pods and ostrich feathers.

Another important note of caution when using air current tubes is that if released in the vicinity of an HVAC smoke detector, it can trigger this device. If in addition to setting off a local alarm, it also brings the local fire department, it can be very embarrassing. Fortunately for me, the one time it happened to me, the connection to the fire department had been disabled for another reason and it was after-hours, so there was hardly anyone else in the building. Let this be a warning to you!

Pressure-Sensing Devices

Pressure-sensing devices are important as part of the IAQ evaluation process because they can indicate several operational characteristics of the ventilation system. Dwyer Instruments of Michigan City, IN manufactures a line of Magnehelic™ differential pressure gauges that can be used for this purpose. The most sensitive of these devices, useful for documenting the pressure differential across the building shell, has a full-scale range of 0.25 in. of water column and costs around $60. Also available are digital micromanometers which are more sensitive, and more expensive, but are very convenient to use.

Flow Measurement Devices

Devices to measure air velocities are useful in performing IAQ evaluations. The basic tools for measuring air velocities in ducts are the combination of a pitot tube and a pressure-sensing device such as an inclined manometer, a Magnehelic™ gauge or a micromanometer. The pitot tube is manufactured of two concentric tubes, the inner one measures the total or impact pressure existing in the air stream, while the outer one measures only the static pressure. Connecting a pressure gauge between these two measurements, the resulting difference between the total pressure and the static pressure is the velocity pressure. The velocity pressure can be used to determine the velocity of the air stream if the density of the air is known. For air at standard conditions, with a density, ρ, equal to 0.075 lb_m/ft^3, the relationship becomes:

$$Velocity = 4005\sqrt{velocity\ pressure}$$

where the velocity pressure term is in inches of water column (in., w.c.) and the velocity term is in feet per minute (fpm). Owing to the difficulty in reading an inclined manometer at an accuracy greater than 0.01 in., w.c., the use of this technique is not recommended for velocities less than 700 fpm.

For measuring low velocities, various types of anemometers exist. These include rotating vane anemometers, hot-wire anemometers, and swinging vane anemometers, also called velometers. Since these are not considered primary standards, a record of calibrations needs to be maintained in order to have confidence in their readings.

For measuring the volume of air at either a supply or exhaust register, a very useful tool is a flowhood device which traps the airflow and measures the velocity pressure across its known cross-sectional area.

Other Useful Tool Box Residents

Tools for gaining access into small AHUs are a requirement for performing IAQ inspections. These tools include a socket and ratchet set, straight-bladed and Philips-head screwdrivers, and a good flashlight. A small mirror mounted on the end of an adjustable rod can be useful for extending the range of visual inspections. A magnifying lens should also be handy. A gallon jug of water is useful for pouring water into a condensate pan to see if the water puddles or drains away. Other useful items include rpm meters or strobes for determining the speed of fans. Amp-meters are also useful for assessing the performance characteristics of fans.

Additional equipment to consider are devices for particle counting, such as a $4000 laser particle counter. Sampling equipment for performing airborne microbiological investigations can also come in very handy. The list can go on and on.

MEASUREMENT OF CARBON DIOXIDE CONCENTRATIONS

The measurement of CO_2 concentrations provides a very powerful technique for evaluating the quantity of outdoor air delivered to occupied spaces in a nonindustrial building. As already mentioned in Chapters 4 and 5, CO_2 measurements can be used to help determine the quantity of outdoor air entering the HVAC equipment *and* to determine the quantity of outdoor air being delivered to building occupants. There are other assessments that the measurement of this component of the indoor air can provide. These other assessments include the adequacy of the duration of ventilation, the impact of parking garages, the impact of combustion sources, and the presence of short-circuiting.

In order to use measurements of CO_2 concentrations as an effective part of an IAQ evaluation, there are many assumptions that must be fully understood in order to prevent inaccurate conclusions. It must be realized that CO_2 concentrations are a dynamic phenomena that will change over time as a function of the duration and number of people present, as well as the operation and efficiency of the ventilation system. It is this dynamic relationship between the CO_2 concentrations and the quantity of outdoor air delivered to the building occupants that needs to be appreciated. I have seen reports where the CO_2 measurements were presented in the form of a time-weighted average over 8 hours. By averaging this data, its potential usefulness is reduced. What is useful to analyze is not the average, but the minimum, maximum, and rate of change in between that characterizes the CO_2 concentrations. The reporting of CO_2 concentrations should also include the time of each measurement, the number of people present, and the duration of their occupancy. The following discussion therefore focuses on understanding the dynamic relationship between the occupancy, ventilation, and CO_2 concentrations.

Relationship between CO_2, Occupancy, and the Ventilation Rate

The relationship between CO_2 concentrations, occupancy, and the ventilation rate is relatively straightforward. Prior to occupancy, the CO_2 concentration in a space should be equal to the outdoor concentration (having been sufficiently ventilated to dilute and remove all air contaminants of human origin from prior occupancy). I have observed a steady increase, from Monday to Friday, of the minimum morning CO_2 concentration from 410 to 530 ppm. After occupancy, the CO_2 concentration will begin to increase in the space, as compared with the outdoors, due to the expired CO_2. This concentration will continue to increase until such time that the absolute quantity of CO_2 leaving the space is equal to that quantity being generated in that space. At that time, steady-state (or equilibrium) conditions will have been achieved and the concentration will stay at a constant value. As discussed in Chapter 5, the assumption that steady-state conditions have been achieved with respect to the build-up of CO_2 concentrations when in fact these conditions have not yet been achieved will lead to overestimates of the amount of ventilation being provided.

Another assumption in the derivation of the relationship between measured peak indoor CO_2 concentrations and effective ventilation rates is that there are no other significant sources of CO_2 present in the space. The reason for the modifying term "significant" as it relates to sources of CO_2 is that concerns have been expressed that the presence of smoking, being a source of CO_2, would confound these results. Calculating the magnitude of this source, in relation to that of people, is illuminating. According to the National Academy of Sciences document "Indoor Pollutants" (National Academy Press, 1981), an emission factor is given for sidestream smoke of 79.5 mg/cigarette over a 550-sec duration of smoke production. This corresponds to 8.7 mg/min of CO_2 production per cigarette. The per-person CO_2 generation rate of 0.3 liters/min corresponds to 456 mg/min. Therefore, the contribution from the cigarette is only 2% that from a person. Since there is already a potential 5% variation in the generation rates expected among individuals in a group, owing to differences in diet, metabolism, and activity level, the presence of the CO_2 from smoking is not significant.

"Significant" sources of CO_2 do include combustion devices. Examples include poorly or inadequately vented gas-fired hot water heaters, cracked heat exchangers in gas- or oil-fired furnaces, or exhaust from automobiles or trucks.

Basic information on the amount of ventilation being provided may be obtained, given an understanding of the operating condition of the HVAC system, by merely making two measurements of CO_2 concentrations, one indoors and one outdoors, at around 3:00 p.m. However, since the benefits of taking more measurements and thereby gaining more information are sufficiently great, it is recommended that the serious practitioner use a portable real-time CO_2 analyzer rather than merely indicator tubes. There is the additional benefit that this analytical equipment can be calibrated, both before and after the on-site building measurements, while the indicator tubes cannot. Some people would consider this a requirement for doing defensible work.

As the field has evolved and developed, one of the most noticeable changes has been the parallel growth of applicable instrumentation. One of the first instruments devoted to the measurement of CO_2 concentrations is the Model RI-411 series, which sells for under $3000 and is available from either GasTech (8445 Central Avenue, Newark, CA 94560-3431, 415-794-6200) or CEA Instruments, Inc. (16 Chestnut Street, Emerson, NJ 07630, 201-967-5660). Since its introduction, this monitor has improved the sensitivity of its LCD display from the nearest 50-ppm concentration to the nearest 25 ppm. GasTech also has its Model 4776 with an analog display and a reported accuracy of ±5% full scale. With a range of 0 to 3000 ppm, this represents an accuracy of 150 ppm. The response of this unit is reported to be 90% of final in 30 seconds. The recent entry into this field is from Gaztech Corporation (6489-A Calle Real, Goleta, CA 93117, 805-964-1699); its product has a much slower response time, on the order of a few minutes. However, at a cost of under $400, this device, which provides measurements to the nearest 1 ppm, appears to be well worth the investment. The challenge with this device is making sure that it is calibrated correctly.

All of these devices need to be calibrated. In terms of calibration, the protocol should use three points, an upper span value of about 1800 ppm, a mid to lower span value of about 400 ppm, and a zero point. The first two points can be checked with the use of prepared calibration gases, while the zero point can be checked with either zero gas such as pure nitrogen or, more conveniently, with a soda lime scrubber. At a minimum, outdoor air can be used as a rough calibration, making sure that this value is between about 330 and 380 ppm for clean areas. Urban areas near highways during a summer inversion can easily exceed 450 ppm outdoors. Readings of 250 ppm should be a clue to an investigator that the device is out of calibration.

The following sections discuss the information that can and should be obtained by extending the number of CO_2 measurements beyond a single set, indoor and outdoor, of peak afternoon values.

Temporal Variations in Carbon Dioxide Concentrations

As discussed previously, it is expected that the CO_2 concentrations will build up through the day to some peak afternoon value in response to the presence of the respired CO_2 from the people in the building. An additional question to be answered, however, is what is the CO_2 concentration the first thing in the morning before the people have started to arrive? The importance of this information is that in a properly functioning building, the early morning CO_2 concentration should be equal to the outdoor concentration because the ventilation system should have purged the building completely of the air contaminants of human origin from the previous day. Remember that CO_2 serves as an indicator for air contaminants of human origin. Therefore, if the early morning indoor CO_2 concentrations are not as low as the outdoor values, then several hypotheses need to be investigated.

One major cause of this situation is that the ventilation system had been shut down too early in the evening with respect to when the people in this building actually leave for the evening. That is, there were still people in the building beyond when the outdoor air was shut off. The other factor necessary in this situation is that the building shell is sufficiently tight for this excess CO_2 to persist overnight.

There are also other possible explanations for elevated indoor CO_2 concentrations in the early morning. Two specific explanations both relate to the possible influence of combustion sources. The two most likely ones are that exhaust gases from cars in a parking lot are being drawn into the building, or that combustion gases from a heating device are entering the building, perhaps via a cracked heat exchanger or due to downdrafting.

CO_2 testing is a very powerful tool for performing IAQ evaluations in occupied buildings. Sometimes, however, a building has yet to have, or no longer has, occupants in it. This is where tracer testing needs to be used. While this tracer can be almost any non-toxic, nonreactive chemical, most researchers have used SF_6. CO_2 can still be used, however, because it can be generated from sources other than people. One potential source is the compressed CO_2 used for carbonating beverages. Starting with an initial concentration of 3500 to 4000 ppm, tracer testing techniques can be performed with CO_2. The following discussion on the use of SF_6 can therefore also be applied to other gases such as (CO_2) as well.

SULFUR HEXAFLUORIDE TRACER TESTING

The information that can be derived from tracer testing techniques can be summarized as follows.

1. Quantification of effective ventilation rates
2. Quantification of distribution inefficiencies in ventilation systems
3. The percentage of outdoor air in the supply air
4. Identification of pathways of air movement through buildings
5. Quantification of reentrainment into air intakes from nearby exhaust stacks
6. Measurements of volumetric air flow rates in ducts
7. Determination of age of air calculations

The use of tracer gas measurements provides a very useful tool in the performance of IAQ evaluations. These evaluations are based on the fact that the achievement of good IAQ is a function of both the delivery of adequate quantities of clean outdoor air to the building occupants *and* the absence of excessive air contaminants in that space. With respect to the quantity of outdoor air, tracer testing can quantify the amount delivered to the occupied areas of the building and also characterize the performance of the ventilation systems. With respect to the presence of air contaminants, tracer testing can identify pathways of air movement that permit odors to be transported to the occupied areas of the building.

Types of Tracer Tests

The concept of an IAQ assessment based on tracer testing relies on the measurement of a single component of the air in a building in order to characterize some aspect of that building's condition. The tracer gas used can either be introduced, as in the case of SF_6, or can already be present, such as the case with some CO_2 based assessments.

The quantification of the amount of outdoor air being delivered to the building occupants, herein referred to as the effective ventilation rate (EVR), can be achieved by one of three tracer gas methods: (1) the concentration-decay method, (2) the constant-emission method, or (3) the constant-concentration method. The most frequently used method of determining the EVR is the concentration-decay method. Another term for this technique is the tracer decay method.

Tracer Decay Method

In this test method, a small quantity of the tracer is released and thoroughly mixed into the space to be evaluated. The method of tracer release used needs to achieve a uniform distribution of tracer within the space. If the ventilation system is creating a well-mixed condition in the building, then the tracer can be released into the return air ductwork, the OA dampers, or the mixing chamber of the AHU. If the building is not well mixed, then the tracer needs to be injected directly into the occupied spaces in amounts that are proportional to the volume of the individual zones involved. After it is confirmed that uniformity throughout the spaces has in fact been achieved, measurements are then obtained to quantify the rate of change of the tracer concentration over time.

As performed by myself, where the individual tracer samples are collected manually into Tedlar air sampling bags with the aid of squeeze bulbs, the information collected includes the sampling location designation, the time of sample collection, and the sampling bag number. I have also started using an automated sampling rig where aquarium pumps are controlled by programmable timers.

The assumption underlying the tracer gas assessment technique to determine ventilation rates is that with perfect mixing with a steady airflow rate, the loss rate of tracer from a space is proportional to its concentration in that space. This relationship is known as the exponential dilution law. As described previously in Chapter 6, this assumption can be mathematically presented as follows.

$$C(t) = C_o \exp^{(-It)} \qquad (6\text{-}4)$$

where

$C(t)$	=	tracer gas concentration at time t
C_o	=	tracer gas concentration at time = 0
I	=	air change rate, per unit time
t	=	time (usually in hours)

Equation 6-4 can be rearranged as follows to solve for the ventilation rate, I:

$$I = \left(\frac{1}{t}\right) \ln\left(\frac{C_o}{C}\right) \qquad (6\text{-}5)$$

For comparing the ventilation rate associated with a change in tracer concentration between two points in time, t_1 and t_2, this equation can be reexpressed as per Equation 6-6.

$$I = \left(\frac{1}{\Delta t_{1 \to 2}}\right) \ln \left(\frac{C_1}{C_2}\right) \qquad (6\text{-}6)$$

As recommended in Chapter 6, a plot of the collected data should be prepared in the form of a semilog graph with the tracer concentration on the vertical log scale and time on the horizontal linear scale. Examples of this type of plot are given in Figure 9.1 for four simple models ranging from perfect mixing to complete nonmixing.[2] In this case, the high amount of recirculation in the single zone fostered the well-mixed case with the resulting straight line plot on a semilog graph.

Tracer Gas Characteristics

In addition to the use of SF_6 as a tracer gas, other gases such as hydrogen, helium, carbon monoxide, carbon dioxide, nitrous oxide, ethane, and methane have been used for the tracer dilution method.[3] The reason that SF_6 is typically selected, however, is because it ranks so highly in having the following basic desirable properties of a tracer gas.

1. The tracer is inexpensive and readily available.
2. It is not adsorbed by walls and furnishings.
3. The analytical method for the gas has negligible cross-sensitivity for other constituents in air.
4. It can be measured accurately at very low concentrations, with inexpensive portable equipment.
5. It has high chemical stability and does not decompose or react with building surfaces or constituents of air.
6. It has no adverse health effects in the concentrations used.
7. It is neither flammable nor explosive.

Testing Requirements

In order to perform the tracer decay method, the equipment required includes a source of the tracer gas, a means for collecting samples of the tracer in the room air, and a tracer gas analyzer. This analysis is usually performed using a gas chromatograph with an electron capture detector. Suitable equipment for this use includes the portable Model 505 Sulfur Hexafluoride Detector/Chromatograph (from Ion Track Instruments, Inc. 340 Fordham Road, Wilmington, MA 01887-2189, 508-658-3767), or the Model 215 AUP portable unit (from Lagus Applied

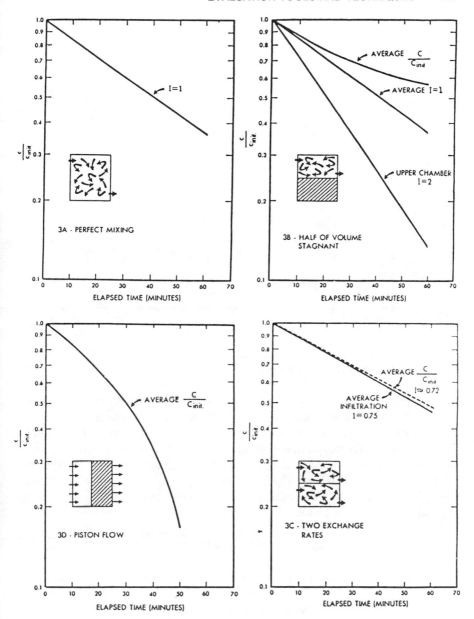

Figure 9.1   Tracer plots for differing mixing characteristics. A, perfect mixing; B, half of volume stangnant; C, piston flow; and D, two exchange rates. From Hunt.[2]

Technology, Inc. 11760 Sorrento Valley Road, Suite M, San Diego, CA 92121, 619-792-9277). Another versatile analytical tool for measuring tracer concentrations is the photoacoustic, multi-gas monitor, Model 1302 from Brüel &

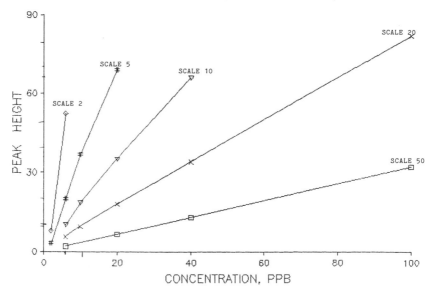

Figure 9.2   Calibration curve for tracer detector.

Kjær Instruments, Inc. (185 Forest Street, Marlborough, MA 01752-3093, 508-481-7000). It needs to be remembered that with analytical devices that rely on the electron capture detector, this detector can become saturated and attempted measurements above this point will deviate from a linear response. Therefore, the linear response range for each piece of analytical equipment being used needs to be determined prior to testing. Figure 9.2 presents a typical calibration curve for this type of equipment. The equipment requirements for performing this procedure were discussed in Chapter 6.

Calculation of Tracer Gas Flow Release Rates

The procedures involved in the actual release of the tracer into the building is very important because of several potential problems. These potential problems relate to the very low sensitivity of the detection equipment and the fact that this detector can be saturated. If too much tracer is released, many hours can be wasted until this concentration in the building has dropped down to the appropriate measurement range. Therefore, the following procedures are recommended. First, all of the tracer release equipment, especially the tank of compressed SF_6, should be stored out of the building to be evaluated until after a check has been made to confirm that there are no background concentrations detectable with the analytical equipment. Then, calculations should be performed to estimate the appropriate tracer release rate. The tracer release set-up (consisting of the tank of compressed gas, regulator, stop valve, needle valve, flow meter, and distribution tubing) can then be adjusted to the required flow rate. Care must be taken so that tracer released during this set-up procedure does not enter the building, either through reentrainment or by adhering to the clothes of the researcher. SF_6 is a very sticky molecule and dealing with part per billion (ppb) quantities of anything requires much forethought to prevent inadvertent contamination.

The equation that can be used to determine the amount of tracer to be released is:

$$F = \frac{V_s \times C_0 \times I}{1 - e^{-It}} \qquad (9\text{-}1)$$

where

F	=	tracer gas flow rate, m³/min
V_s	=	effective volume of the structure, m³
C_0	=	concentration of tracer to be achieved
I	=	ventilation rate, air changes per hour (ACH)
t	=	time of the charging period, hours

Note that the concentration value is expressed as a dimensionless fractional value, where 1 ppm is equal to 0.000001 or 1.0×10^{-6}. Similarly, the initial tracer concentration of 50 ppb (that I aim for) is expressed in this equation as 5.0×10^{-8}. In Equation 9-1, the term in the denominator, $(1 - e^{-It})$, eventually goes to unity because the (e^{-It}) term goes to zero as the product of the time, t, and the

Table 9.1.   Values for the Time Constant $(1 - e^{-It})$, As a Function of Ventilation Rate and Time

Ventila-tion rate, ACH of OA	Time Elapsed, Minutes and Hours									
	10 0.17	20 0.33	30 0.50	40 0.67	50 0.83	60 1.00	70 1.17	80 1.33	90 1.50	100 1.67
0.2	0.033	0.064	0.095	0.125	0.154	0.181	0.208	0.234	0.259	0.283
0.3	0.049	0.095	0.139	0.181	0.221	0.259	0.295	0.330	0.362	0.393
0.4	0.064	0.125	0.181	0.234	0.283	0.330	0.373	0.413	0.451	0.487
0.5	0.080	0.154	0.221	0.283	0.341	0.393	0.442	0.487	0.528	0.565
0.6	0.095	0.181	0.259	0.330	0.393	0.451	0.503	0.551	0.593	0.632
0.7	0.110	0.208	0.295	0.373	0.442	0.503	0.558	0.607	0.650	0.689
0.8	0.125	0.234	0.330	0.413	0.487	0.551	0.607	0.656	0.699	0.736
0.9	0.139	0.259	0.362	0.451	0.528	0.593	0.650	0.699	0.741	0.777
1.0	0.154	0.283	0.393	0.487	0.565	0.632	0.689	0.736	0.777	0.811
1.1	0.168	0.307	0.423	0.520	0.600	0.667	0.723	0.769	0.808	0.840
1.2	0.181	0.330	0.451	0.551	0.632	0.699	0.753	0.798	0.835	0.865
1.3	0.195	0.352	0.478	0.580	0.662	0.727	0.781	0.823	0.858	0.885
1.4	0.208	0.373	0.503	0.607	0.689	0.753	0.805	0.845	0.878	0.903
1.5	0.221	0.393	0.528	0.632	0.713	0.777	0.826	0.865	0.895	0.918

Table 9.2.   Tracer Release Rates (in cc/min) to Achieve Desired 50 ppb Concentration in 18,000 m³ Volume for Various Ventilation Rates and Charging Times

Ventilation rate, ACH of OA	Time Elapsed, Minutes and Hours									
	10	20	30	40	50	60	70	80	90	100
	0.17	0.33	0.50	0.67	0.83	1.00	1.17	1.33	1.50	1.67
0.2	91.5	46.5	31.5	24.0	19.5	16.5	14.4	12.8	11.6	10.6
0.3	92.3	47.3	32.3	24.8	20.3	17.4	15.2	13.6	12.4	11.4
0.4	93.0	48.1	33.1	25.6	21.2	18.2	16.1	14.5	13.3	12.3
0.5	93.8	48.9	33.9	[26.5]	22.0	19.1	17.0	15.4	14.2	13.3
0.6	94.6	49.6	34.7	27.3	22.9	19.9	17.9	16.3	15.2	14.2
0.7	95.4	50.5	35.6	28.2	23.8	20.9	18.8	17.3	16.2	15.2
0.8	96.1	51.3	36.4	29.0	24.7	21.8	19.8	18.3	17.2	16.3
0.9	96.9	52.1	37.3	29.9	25.6	22.7	20.8	19.3	18.2	17.4
1.0	97.7	52.9	38.1	30.8	26.5	23.7	21.8	20.4	19.3	18.5
1.1	98.5	53.8	39.0	31.7	27.5	24.7	22.8	21.4	20.4	19.6
1.2	99.3	54.6	39.9	32.7	28.5	25.8	23.9	22.6	21.6	20.8
1.3	100.1	55.5	40.8	33.6	29.5	26.8	25.0	23.7	22.7	22.0
1.4	100.9	56.3	41.7	34.6	30.5	27.9	26.1	24.8	23.9	23.3
1.5	101.7	57.2	42.6	35.6	31.5	29.0	27.2	26.0	25.2	24.5

ventilation rate, I, increases to infinity. The closer this "time constant" term is to unity, the closer the conditions are to steady-state. It may be interesting to the reader to observe typical values for this term as a function of ventilation rate and time. This information is presented in Table 9.1.

One observation that can be made from a review of Table 9.1 is that for a ventilation rate of 1.4 ACH and a time interval of 100 min, the tracer build-up will be at 90% of its ultimate peak steady-state value.

Another observation is that for 1.0 ACH and a time interval of 1 hour, the build-up will be to 63.2% of its steady-state concentration. This value also represents the amount of air that will be displaced for the well-mixed case.

Example 9.1. In a recent building evaluation, tracer was to be released into a building with a reported total area of 54,300 ft² and measured total floor-to-ceiling height of 13 ft. This corresponds to a total gross interior volume of 705,900 ft³. Recognizing that not all of this interior volume is available for the tracer to disperse into, due to the presence of furnishings and stagnant areas, the effective volume was estimated to be 90% of this gross value, or 635,310 ft³. Converting this value to cubic meters (by multiplying it by 0.02832) yields 17,992 m³. For ease of calculation purposes, this value is rounded up to 18,000 m³. Based on Equation 9-1, the tracer release rates to be used to achieve a concentration of 50 ppb for various combinations of ventilation rates and charging times can be calculated. To convert the tracer release rate from cubic meters per minute to cubic centimeters per minute, the value calculated from Equation 9-1 is multiplied by 1,000,000. These values are presented in Table 9.2 for ventilation rates ranging from 0.2 to 1.5 ACH and charging times ranging from 10 to 100 min.

Table 9.3.   Total Quantity of Tracer Released (in cc) to Achieve 50 ppb in
             18,000 m³

Ventila-tion rate, ACH of OA	Time Elapsed, Minutes and Hours									
	10 0.17	20 0.33	30 0.50	40 0.67	50 0.83	60 1.00	70 1.17	80 1.33	90 1.50	100 1.67
0.2	915	930	946	961	977	993	1009	1025	1042	1058
0.3	923	946	969	993	1017	1042	1067	1092	1118	1144
0.4	930	961	993	1025	1058	1092	1126	1161	1197	1233
0.5	938	977	1017	1058	1100	1144	1188	1233	1279	1326
0.6	946	993	1042	1092	1144	1197	1251	1307	1365	1424
0.7	954	1009	1067	1126	1188	1251	1317	1384	1454	1525
0.8	961	1025	1092	1161	1233	1307	1384	1464	1545	1630
0.9	969	1042	1118	1197	1279	1365	1454	1545	1640	1738
1.0	977	1058	1144	1233	1326	1424	1525	1630	1738	1849
1.1	985	1075	1170	1270	1375	1484	1598	1716	1838	1964
1.2	993	1092	1197	1307	1424	1545	1672	1804	1941	2082
1.3	1001	1109	1224	1346	1474	1608	1749	1895	2046	2202
1.4	1009	1126	1251	1384	1525	1672	1827	1987	2154	2326
1.5	1017	1144	1279	1424	1577	1738	1906	2082	2264	2451

In this particular building, since I estimated a ventilation rate of around 0.5 ACH and I wanted a charging interval of 40 min, this led to the selection of a tracer release rate of 26.5 cc/min.

For this example, the total amounts of tracer that would be released as function of ventilation rate and tracer release interval are presented in Table 9.3.

For the volume of 18,000 m³, the amount of tracer that would need to be present to achieve an instantaneous concentration of 50 ppb would be equal to 900 cc. That is, if there were a sealed box with this volume and 900 cc of tracer were released and mixed into it, the resulting concentration would be 50 ppb. However, if this box is being ventilated, then the amount of tracer introduced will need to be greater than 900 cc because some of the tracer will have been removed during the tracer release interval. As can be observed in Table 9.3, the greater the ventilation rate and the longer the tracer release interval, the greater the amount of tracer that needs to be released. For instance, with a ventilation rate of 1.2 ACH and a release interval of 80 min, half of the tracer released will have been removed from the building at the end of the release interval.

Advantages and Disadvantages

Tracer decay measurements are a powerful tool for evaluating the performance of ventilation systems. One advantage of this approach is the amount of information that can be obtained per unit time allocated to testing. For a well-mixed space being provided with a ventilation rate of 1.0 ACH, the halving time for the tracer concentration is 41.58 min. Halving times are mentioned because they give an estimate of the minimum necessary duration of time required for a tracer sam-

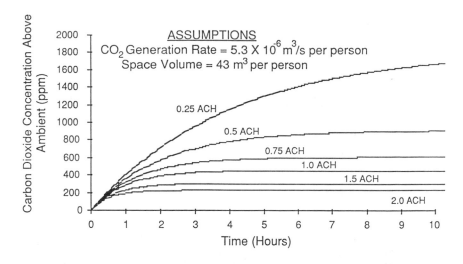

Figure 9.3   Time required to achieve equilibrium conditions. (From Persily, A. and W. S. Dols. 1990. *Air Change Rate and Air-Tightness in Buildings, ASTM STP10067.* Sherman, M. H., Ed. American Society for Testing and Materials, Philadelphia. pp. 77–92. With permission.)

pling. For a well-mixed space being provided with a ventilation rate of 0.5 ACH, the halving time for the tracer concentration is 83.2 min. Therefore, the time interval that the HVAC and building parameters need to be held constant is in the range of 1 to 2 hours. This contrasts favorably with the use of CO_2 measurements which typically need too many hours for equilibrium conditions to be achieved. For instance, at a ventilation rate of 1.0 ACH, it would take almost 4 hours for the build-up of CO_2 concentrations to reach equilibrium values. The time required to achieve equilibrium conditions and other ventilation rates and presented in Figure 9.3. This issue was discussed in detail in Chapter 6.

Another advantage of SF_6 tracer testing, as compared with CO_2 testing, is also related to the reduced time for obtaining useful information. That is, the presence of the tracer typically persists in the building long enough that the parameters of the HVAC system can be intentionally adjusted during the decay to provide additional information. For instance, the relationship between differing positions of the OA dampers and the resulting effective ventilation rates can be quantified. The results of such a test are presented in Figure 9.4, where for the AC-9 zone the EVR changed from 0.5 ACH to 1.2 ACH.

Distribution Inefficiencies

As part of the evaluation of the adequacy of the quantity of outdoor air being delivered to building occupants, there are basically two determinations:

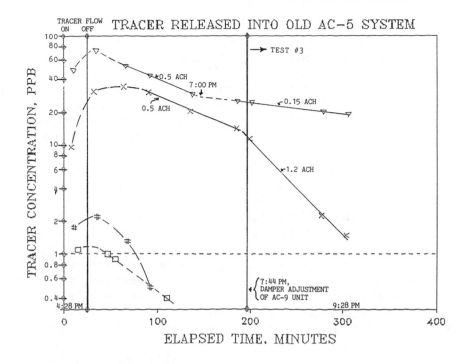

Figure 9.4 Tracer test results for two outdoor air damper positions.

1. How much outdoor air is entering the HVAC system?
2. How much of this outdoor air is actually being distributed to the building occupants?

Therefore, for a given building, although a potentially adequate quantity of outdoor air is entering the HVAC system, distribution inefficiencies may prevent an adequate quantity of outdoor air from being delivered to the building occupants. Distribution inefficiencies can be categorized into three basic types: (1) leakage of supply air into the return plenum, (2) short circuiting of supply air at the ceiling, and (3) delivery of supply air to unoccupied areas.

The comparison of tracer concentrations in conjunction with, or separate from, ventilation rate determinations can be used to evaluate the presence or absence of these conditions.

To assess if leakage is occurring from the SA ductwork to the return plenum, there are two sampling protocols that can be followed. The first approach involves a slow release of tracer into the OA intake with comparative tracer measurements being obtained at the following locations: several SA registers, representative occupied space locations, and the return air to the AHU. If leakage is occurring, the tracer build-up in the return air to the AHU will precede the build-up at the in-space locations. The requirement for several SA registers is included to make sure that the tracer is being uniformly distributed throughout the building, an important consideration for meaningful results. The SA registers selected should therefore represent the differing major branches of the supply air distribution sy.

The second approach for assessing duct leakage involves obtaining comparative tracer measurements after the tracer release has stopped. In this situation, the only air tracer concentration needs to be lower than the in-space concentration. If leakage is occurring, the tracer concentration in the return air to the AHU will drop more rapidly than the concentrations measured at the in-space locations.

The other category of distribution inefficiencies, that of short-circuiting occurring in the occupied space as opposed to in the return air plenum, can be assessed with a similar technique. This time, there needs to be more in-space measurement locations that vary not only in their distribution throughout the space, but also in their height off the floor. Specifically, both the build-up and decay of tracer concentrations should be compared for locations near the ceiling with those at the breathing height for seated individuals. If this short-circuiting is occurring, then the ceiling location will first display, during the release portion of the test, a more rapid build-up of the tracer as compared with the breathing zone location. Conversely, during the decay portion of the test, the ceiling location will exhibit a more rapid decrease in tracer concentration than for the breathing zone location. Care therefore needs to be exercised that sampling begins before the tracer release has terminated in order to assess the magnitude of any differences. The complexity of the mathematics of this approach typically limit the determinations to qualitative results. Readers interested in pursuing the theory and mathematics necessary to apply the tracer gas method to buildings of many chambers are referred to the work of Sinden[4] on multi-chamber theory of air infiltration.

Percentage of Outdoor Air

The determination of percentage of outdoor air (% OA) in the supply air is important for several reasons. It can be used as part of the determination of the quantity of outdoor air entering the HVAC equipment. It can also be used as part of another technique for determining the quantity of outdoor air delivered to building occupants. In this approach, the quantity of supply air serving a given area is determined by measuring the flow quantity at each supply diffuser and then this quantity is multiplied by the % OA in the supply air to yield the amount of outdoor air being delivered to that area. While the quantity of supply air is typically determined by using a flowhood device, the % OA can be determined by a number of techniques, including the use of tracer techniques. This technique was described in detail in Chapter 5.

Pathways of Air Movement

The determination of pathways of air movement through buildings is an important part of IAQ evaluations because odors and other air contaminants can be transported from one location to another along these routes. Tracer gas testing can be used to both identify the existence of these pathways and the specific location routes that they take. Unlike a determination of effective ventilation rates, which can be considered a basic component of practically all IAQ evaluations, a determination of the pathways of air movement will typically only be performed if there is a specific hypothesis to be tested. Typical reasons include complaints of odors from loading docks, dumpsters, sewer gas, parking garages, or animal care facilities. These investigations were summarized in Chapter 8.

Reentrainment Quantification

As mentioned in Chapter 5 as part of the discussion of the determination of the % OA entering the AHU, the measurement of the tracer concentration in the outdoor air can be used to quantify the amount of reentrainment of general building exhaust occurring. This same approach can also be used to determine if reentrainment into a building's air intakes is occurring from specific nearby exhaust stacks. These stacks are typically from sources that include kitchen, laboratory fume hood, and process exhausts. In this evaluation procedure, it is recommended that a computer modeling effort first be performed to estimate the dimensions of the building's recirculation cavities and to estimate the expected minimum dilution amounts and the expected worst-case wind speeds and directions. Then, during suitable meteorological conditions, a tracer gas testing effort can be performed to validate the results from the computer modeling effort.

For this actual test procedure, the first step is to confirm that background concentrations of tracer are below the detection limit of the analytical equipment. After that is accomplished, a controlled release of tracer is initiated at the identified source of air contaminants, typically a laboratory fume hood. This controlled released involves a source tank of SF_6 gas, a regulator, a needle valve, and a calibrated rotameter. Sampling then begins at the stack head and at the OA intake of concern. Experience has shown the benefit of prediluting the sample collected at the stackhead with tracer-free upwind air. This can prevent saturating the tracer gas equipment and eliminating the need for cumbersome post-sampling, preanalysis dilution techniques. Using calibrated equipment, this technique can quantify both the volume of air being discharged in the exhaust system, as well as the amount of dilution occurring between the exhaust stack and nearby air intakes.

Example 9.2. A 100-cc/min tracer sample is being released into one laboratory fume hood, which is ganged together with five other hoods to one common exhaust fan and stack. Air samples collected at the stackhead are prediluted with clean upwind air, with the ratio of stack air to clean air being 1:10. Sampling is accomplished with a squeeze bulb and a Tedlar air sampling bag with a closable valve. The dilution is achieved by 36 strokes of the squeeze bulb with clean air and 4 strokes at the stackhead. The average of three samples collected at this location yields a concentration of 54.3 ppb. Due to the predilution, the actual stack concentration would therefore be 543 ppb. Multiplying the release rate of 100 cc/min by 3.531×10^{-5} converts this value to 3.531×10^{-3} cfm. The volume of air in the exhaust system can then be calculated as shown in Equation 9-2:

$$\frac{3.531 \times 10^{-3} \text{cfm}}{Q, \text{ cfm}} = \frac{543}{10^9} \qquad (9\text{-}2)$$

where Q is the airflow rate in the exhaust, in cfm. This equation can then be solved for the volumetric flow rate, Q, as shown in Equation 9-3:

$$Q = \frac{3.531 \times 10^{-3} \text{cfm} \times 10^9}{543} = \frac{3.531 \times 10^6 \text{cfm}}{543} = 6503 \text{ cfm} \qquad (9\text{-}3)$$

The air samples collected at a nearby air intake display considerable variation due to varying wind conditions. The three highest values out of the 30 collected — at 2.9, 2.8, and 2.4 ppb — yield an average of 2.7 ppb. Dividing the stack concentration of 543 ppb by the air intake concentration of 2.7 ppb yields the result that a minimum amount of dilution occurring during these test conditions was only 200-fold.

Measurement of Volumetric Flow Rates

In addition to measuring airflow rates in exhaust systems, tracer gas techniques can be used to quantify air flow rates in HVAC ducts. Measuring airflow rates in ductwork in building ventilation systems can be difficult using airflow rate measurement techniques (e.g., pitot tubes) due to access problems and insufficient lengths of straight ductwork. Long lengths of straight ductwork are required for the establishment of uniform flow profiles of in-duct velocities, a prerequisite for accurate measurements. There are two basic approaches to volumetric flow rate determinations in ducts: a continuous release and a pulse release.

Continuous Release Measurement of Airflow Rates

For the measurement of volumetric airflow rates using a continuous tracer release, as it was for the exhaust system test, there is a continuous low-level tracer release of known quantity with repeated air samples collected to determine the instantaneous concentrations downstream. The mathematics involved determine the volume of air that would be required to dilute the known release amount down to the measured concentration. One requirement for the use of this technique is that the tracer concentration needs to be well mixed across the duct section where the sampling is occurring. Since several measurements should be collected anyway as a good analytical procedure, if any variation is detected, then the sampling protocol can involve dividing the duct section into several zones of equal area and collecting equal air quantities from each zone.

Example 9.3. A 20 cc/min tracer sample is being released into the OA intake of an AHU. Air samples collected from the last supply diffusers of the four main supply trunks yield tracer concentrations of 60 ppb to within 10% of each other. Multiplying the release rate of 20 cc/min by 3.531×10^{-5} converts this value to 7.062×10^{-4} cfm. The volume of air in the system can then be calculated as shown in Equation 9-4.

$$\frac{7.062 \times 10^{-4} \text{ cfm}}{Q, \text{ cfm}} = \frac{60}{10^9} \tag{9-4}$$

where Q is the airflow rate in the duct, in cfm.
    This equation can then be solved for the volumetric flow rate, Q, as shown in Equations 9-5:

$$Q = \frac{7.062 \times 10^{-4} \text{ cfm} \times 10^9}{60} = \frac{7.062 \times 10^5 \text{ cfm}}{60} = 11,770 \text{ cfm} \tag{9-5}$$

When attempting to measure low volumes of airflow, this technique may prove impractical due limitations in being able to release small enough quantities of tracer without saturating the detector. In these situations then, the pulse-injection tracer techniques can be used.

Pulse Release Measurement of Airflow Rates

The duct pulse technique is described in detail in by Persily and Axley.[5] In the tracer pulse technique, as with the continuous release approach, the tracer concentration across the duct at the sampling location should be uniform. The air flow rate itself also needs to remain constant. As presented by Persily, the mathematics involved with this approach involve the "application of integral mass balance equations to the reduction of the measured concentration response data." That is, the air mass flow rate is equal to the ratio of the mass of tracer injected to the integral of the concentration response downstream from the injection point. This concentration response will go from zero, up to a peak value, and then tail off back to zero again. Taking the integral of this concentration response curve refers to estimating the area under this curve.

This is much more straightforward than it sounds. The concentration integral is simply the average concentration multiplied by the length of time over which the sample was collected.

Example 9.4. Sampling is begun downstream in the ductwork and will continue for 5 min. Shortly after the start of this 5-min sampling interval, a small quantity of tracer (40 cc, 0.00141 ft^3), is injected. The sampling which began before the injection and continued for exactly 5 min yielded an average value of 28.25 ppb tracer. Expressed as a simple fraction, this concentration of 28.25 ppb is equal to 2.825×10^{-8}. Multiplied by the 5 min, this yields a value of 1.41×10^{-7} minutes. Dividing the injection quantity of 1.41×10^{-3} ft^3 by the value of 1.41×10^{-7} min yields the airflow value of 1×10^4 cfm (or 10,000 cfm).

In applying the duct pulse technique, there are several practical considerations. The most important issues are precisely knowing the amount of the tracer that is injected, the length of the sampling interval, and the average concentration of the integrated sample. This average must be based on a cross-sectional average concentration; that is, the concentration at the point of measurement must be varying only along the length of the duct, not along the duct cross-section. A multi-point injection across the duct cross-section is therefore useful in achieving a uniform concentration at the measurement point.

In applying this technique to a particular system, there will be some initial uncertainty in the amount of tracer to release and the appropriate time of sampling. The primary requirement is that the average concentration in the collected sample be within the calibrated range of the measurement equipment. Estimates of the airflow rate, based on the duct diameter and pitot tube measurements, can be used to predict the tracer release quantity and sampling duration. Since each measurement requires only a few minutes, it should not be difficult to repeat the procedure several times until the appropriate parameters are achieved. Repeating the procedure is also recommended to establish a statistical validity to the results.

Determination of Age of Air Calculations

The age of air in a room is a measure of the length of time it has been in the room. The "youngest" air is found where the supply air, containing the outdoor air, first comes into the room. The "oldest" air, however, may be found anywhere in the room, depending on the location of stagnation zones. Age of air determinations are evolving as a tool to aid in the evaluation of air change efficiency. In this usage, air change efficiency refers to how effectively the air present in a room is replaced by fresh air from the ventilation system. This definition is contrasted with the term "ventilation efficiency" which refers to how quickly a contaminant is removed from the room. Both of these terms are discussed in more detail in Chapter 7 and a publication of the Air Infiltration and Ventilation Center.[6]

There are two different ways of considering the age of air: the *local mean* age of air and the *room average* age of air. The *local mean* ages at specific points in a room can be compared to give a measure of the spatial variations of air distribution within that room. Uses for the local mean age of air include evaluating such situations as individual work stations, naturally ventilated buildings, or in the mapping of airflows through rooms. The big advantage of this method is that the results apply to individual points within a room. This lets areas of stagnant air be located, and the ventilation air supply at head height at an individual's work station can, for example, be assessed.

The basic equations for local mean age of air can be summarized as follows:

local mean age, $\bar{\tau}$

$$\text{1. pulse injection } \bar{\tau}_p = \frac{\int_0^\infty \tau \cdot C(\tau)\, d\tau}{\int_0^\infty C(\tau)\, d\tau}$$

$$\text{2. step - up method } \bar{\tau}_p = \int_0^\infty \left[1 - \frac{C(\tau)}{C(\infty)} \right] d\tau$$

$$\text{3. decay method } \bar{\tau}_p = \frac{\int_0^\infty C(\tau)\, d\tau}{C(0)}$$

where

$\bar{\tau}_p$	=	local mean age of air at point p
$C(\tau)$	=	concentration of tracer at time t
t	=	time
$C(0)$	=	concentration at t = 0

The *room average* age of air is a number which quantifies the performance of a ventilation system. This number takes into account both the amount of ventilation air supplied to the room and the efficiency with which it is distributed around the room. Room average age of air is measured in the exhaust air leaving the room. If there are multiple exhausts, however, then both the tracer gas concentration measurements and the volumetric flow rates of air in each exhaust duct must be known in order to make this determination. The concentrations are then volume weighted in the calculation of the room average age of air determination. These data requirements therefore tend to limit the determinations to research projects or rooms with just one or a few supply and exhaust registers. Typical office spaces have so many exhaust locations as to render this approach impractical.

Another requirement for the use of age of air determinations is that, for all practical purposes, all of the air entering and leaving the room does so via designated supply and exhaust ducts. That is, infiltration and exfiltration are minimal.

REFERENCES

1. *ASHRAE Standard 62-1989.* Appendix D.
2. Hunt, C. M. 1980 "Air Infiltration: A Review of Some Existing Measurement Techniques and Data," *Building Air Change Rate and Infiltration Measurements. ASTM STP 719,* C. M. Hunt, J. C. King, and H. R. Trechsel, Eds., American Society for Testing and Materials, Philadelphia, pp. 3–23.
3. American Society for Testing and Materials. 1983. *Standard Test Method E741-83, Determining Air Leakage Rate by Tracer Dilution.* ASTM, Philadelphia, 12 pp.
4. Sinden, Frank W. 1978. "Multi-Chamber Theory of Air Infiltration" *Building and Environment,* Vol. 13. Pergamon Press, Elsmford, NY. pp. 21–28.
5. Persily, A. and J. Axley. 1990. "Measuring Airflow Rates with Pulse Tracer Techniques," *Air Change Rate and Airtightness in Buildings, ASTM STP 1067,* M. H. Sherman, Ed., American Society for Testing and Materials, Philadelphia, pp. 31–51.
6. Air Infiltration and Ventilation Center (AIVC) Technical Note 28. "A Guide to Air Change Efficiency." AIVC, University of Warwick Science Park, Barclays Venture Centre, Coventry, England, 32 pp.

CHAPTER 10

Sources of Air Contaminants

OVERVIEW

This chapter presents a discussion of the potential sources of air contaminants other than those arising merely from the occupants. This discussion is included since indoor air quality (IAQ) problems exist when there is an inadequate delivery of sufficient quantities of clean outdoor air for ventilation in comparison with the quantity of air contaminants that are present in occupied areas. This is a dynamic relationship between the amount of ventilation required and the air contaminants present: the more contaminants present, the greater the amount of ventilation required to achieve a certain level of IAQ.

In terms of the HVAC system and air contaminants, there are three types of conditions that need to be considered:

1. The HVAC system may be a source of air contaminants.
2. The HVAC system may be functioning as a transfer mechanism to transport air contaminants from their source to where the people are.
3. The HVAC system may be failing to adequately dilute and remove the air contaminants that are arising from a source that is located in the occupied space.

Each of these situations is addressed in this chapter. Also included in this chapter is a discussion of the various categories of sources of air contaminants, as well as a description of the steps necessary to evaluate the role of the ventilation system with respect to these air contaminants.

Table 10.1.   Sources of Air Contaminants by Location of Origin

Air contaminants from outdoors via outdoor air intake
Air contaminants from outdoors via unintentional intakes
Air contaminants arising from within the HVAC system
Air contaminants arising from the building itself

There are several approaches for characterizing sources of indoor air contaminants. Three examples include organizing them according (1) to their location of origin, (2) to the activity that generates them, or (3) to the types of air contaminants and their properties.

SOURCES OF AIR CONTAMINANTS BY LOCATION OF ORIGIN

Sources of air contaminants can arise in several locations. These potential sources of air contaminants and their relationship to the evaluation of the adequacy of the ventilation systems are summarized in the categories listed in Table 10.1.

Air Contaminants Arising from Outdoors

As presented in Table 10.1, sources of air contaminants that originate outdoors can enter the building in more than one way. There are, in fact, three basic mechanisms by which air contaminants originating outdoors can be transported to the occupied areas of a building:

1. Air contaminants can be transported via the HVAC system by being captured by the outdoor air intake at the air handling units (AHUs).
2. Air contaminants can be transported via the HVAC system by being captured via less obvious pathways into the HVAC system, such as via the mechanical room or somehow into the return flow to an AHU.
3. Air contaminants can be transported directly to an occupied area by infiltration through the building shell.

Air Contaminants Entering at the Outdoor Air Intake

For air contaminants being drawn into a building at the designated air intake to the AHU, there are several examples that can be mentioned. If the air intakes are situated near the top of the building, the sources to be concerned about include cooling towers, plumbing vents, stagnant roof puddles, and building exhausts, such as from kitchens or laboratory fume hoods. For lower-level air intakes, this list of potential sources includes vehicle emissions from roadways, parking lots, garages, loading docks, dumpsters, trash compactors, etc. For below-grade air intakes, those located in areaways, there is also the potential for microbial contamination from decaying leaf litter or even pigeon droppings.

Some sources outside the building will be a potential threat for air intakes at any location. Nearby construction activities can also be a cause for concern. For instance, excavation of the earth has been implicated in the release of large quantities of legionella bacteria. Other nearby activities, such as industrial operations, can also be a source of pollutants. Nature itself, in season, can also be a source of pollen, dust, and fungal spores.

Air Contaminants Entering via Unintentional Air Intakes

In addition to outdoor sources being transported into the building via the obvious, designated air intakes, there is also the possibility that air contaminants are being drawn into the building via unintentional air intakes. Examples of this situation can involve sources that are located in or near the mechanical rooms themselves. As mentioned previously, in the discussion on establishing pathways of air movement, there were two examples of this situation given. One example had a mechanical room built on hollow, prestressed concrete planks directly above the loading dock, while the other was merely in the basement, but under significant negative pressure with respect to the outdoors, including the loading dock.

Air Contaminants Entering via Infiltration

In addition to outdoor air being drawn into the HVAC system, contaminated outdoor air can also be drawn into the building via an unintentional pathway due to infiltration. The driving force for this infiltration can either be due to the stack effect or pressure differences created by the operation of the HVAC equipment. These situations were discussed in Chapter 8. Depending on where the infiltration occurs, the extent of the impact of any air contaminants it introduces into the building can vary from just a localized area to an entire zone of the building. This area of impact will depend on whether the infiltration is occurring adjacent to an occupied area or adjacent to a mechanical equipment area where the air contaminants can be drawn into AHUs and then distributed.

In situations where there is a parking garage at the base of a building, either below ground or not, it is prudent to check to make sure that air from this garage does not seep into the building. There was recently a problem in a building where the floor of an underground parking garage had a sealer applied and the building above filled up with the fumes from this application. Prior testing with a non-toxic tracer such as sulfur hexafluoride, as discussed in Chapter 8, could have identified both the magnitude of this air transfer and the specific locations involved in this pathway of air transfer. Armed with this information, the problem could have been avoided by several techniques. If the amount of leakage into the building could not be reduced to an acceptably small amount, as determined by tracer testing, extra ventilation could have been added during the application of the sealant to depressurize the garage with respect to the building.

Soil gas sucked into a basement due to negative pressures there can introduce such sources of air contaminants as radon, pesticides, or volatile organic chemicals (VOCs) that have been spilled into the ground from previous site uses or nearby industrial activities.

Air Contaminants Arising from within the HVAC System

Air pollutants can arise from sources within the equipment and ductwork of the HVAC systems itself. There are several locations within that are particularly susceptible to being sources of air contaminants. These locations start with the outdoor air intake chamber itself and continue through the system to include the return air plenum.

Outdoor Air Intake Chamber

The outdoor air intake chamber can be a source of air contaminants if this location is used for the storage of chemicals. Also, as mentioned previously, if there are gaps in the sides of this chamber, air contaminants present in or near the mechanical room can be drawn into the HVAC system at this location. Also, any drains in this location should be checked to make sure that they are not able to permit sewer odors entrance to the AHU. These drains are present so that any rain or snow entering this area will not accumulate.

Air Filters

The air filters can also be a source of odor generation, with the collected particles being the actual source. In particular, an investigation by Hujanen et al.[1] studied the generation of odors by evaluating filters that were collected from office buildings at the end of useful service life, when they were replaced with new filters. This evaluation therefore represented a worst-case scenario for these filters. This study showed that a trained panel was able to quantify the odor generation of the filters. The results showed that not only the dust accumulation on the filters, but also the burden of VOCs on the building must be considered when determining the replacement period of filters.

Condensate Drain Pan

Downstream from the filters, the next location of concern as a potential source of air contaminants is, of course, the condensate drain pan under the coiling coils. Here, the presence of excess moisture or even standing water can promote microbial growth. In terms of preventive maintenance, this location should be inspected at a regular frequency. According to the Air-Conditioning and Refrigeration Institute (ARI),[2] for instance, condensate drain pans should be checked for algae and mold growth every 3 months. Also at this location, the use of biocides or cleaning compounds can also represent a source of air contaminants.

Heat Exchangers

Moving on from the cooling coils to the heating coils, there can be the improper or inadequate venting of combustion products. For heating systems with a heat exchanger directly in the supply air stream, cracks in the exchanger can introduce products of combustion directly into this air stream. If this is occurring, carbon monoxide or carbon dioxide measurements can be used to detect this source of air contaminants.

Supply Air Fan

Continuing through the AHU, the fan itself can also become covered with microbial growth due to the combination of the high humidities downstream of the cooling coils and accumulated dirt that has gotten by the filters and adheres to this equipment.

Internal Duct Liners

Potential microbial growth problems also exist downstream of the fan as well, especially if the ductwork is lined on the inside with porous insulation. Morey and Williams[3] have investigated this issue and have come to the conclusion that "porous insulation that is both dirty and moist or wet can become a strong, primary emission source of indoor microbial pollutants. Porous insulation near AHU drain pans and cooling coils, in air supply ducts immediately downstream of the cooling coil plenum, and on the inside housing of unit ventilators, fan-coil units, and induction units is most likely to become a strong microbial amplification site. The continued use of porous insulation in HVAC systems appears to be incompatible, from a microbiological point of view, with the concept of a healthy building. Accordingly, insulation should be applied outside main air supply ducts and outside other HVAC components that periodically become moist. These installations will require a vapor barrier on the inside of the insulation. This will help in preventing growth of microorganisms and their release into the ventilation air stream. Discouraging the use of porous insulation inside HVAC system components will have a profound effect on the manner in the [way] HVAC systems are designed and constructed in the future." In addition to discouraging the use of porous insulation inside HVAC components, as is already happening in the hospital sector, upgrading the efficiency of filter banks will help slow the rate of accumulation of dirt in ductwork. Besides the accumulation of dirt in the system, the other contributing factor — moisture — cannot be eliminated for any system that cools its supply air to below its dew point.

Another potential problem to be aware of with respect to internal duct liners is that they can become a source of fiberglass particles in the delivered supply air. In highly turbulent locations, such as in the vicinity of the supply air fan, the protective covering over the fiberglass can be eroded away and then the underlining fiberglass fibers will be eroded away as well. These fiberglass particles are not only a respiratory irritant, but when clumps of this material shows up on people's desks, it can rapidly precipitate complaints.

Other HVAC Sources

In addition to the specific components of the HVAC systems, there are also other building mechanical areas that can become sources of air contaminants. The electric motors or the hydraulic fluid for the elevators, for instance, can be a source of air contaminants. Also to be considered as a potential source is the return air plenum. This location can be a reservoir of dust which can be introduced into the occupied areas by fan-powered terminal boxes which use plenum air to supplement the supply air coming from the centralized HVAC equipment. If there are water leaks in this location, either from equipment or from the outdoors, this dust in the presence of moisture can then become a nutrient base and ultimately a source of microbiological amplification.

Air Contaminants Arising from Activities in the Building

For air contaminants originating from within the building, as opposed to outdoors from it, one can categorize these by the activity that generates them. Activities occurring in the building with the potential to generate air contaminates are listed in Table 10.2.

Air Contaminants from the Occupants Themselves

The occupants of buildings themselves represent sources of air contaminants. These air contaminants can be grouped into two basic categories: bioeffluents and "germs." The term "bioeffluent" is a fancy word for the chemicals emitted by people. These include the body odors that are created by the interaction of bacteria living on our skin with the perspiration and dirt present there. In some situations, such as in a respiratory hospital, visitor's wearing perfumes can represent a source of air contaminants. Exhaled carbon dioxide (CO_2) is also considered a bioeffluent. The levels of CO_2 that are typically present in office buildings have not been shown to elicit symptoms of sick building syndrome, or even occupant complaint syndrome. The measurement of CO_2 levels is nevertheless important for several reasons. First of all, it is reasonable to assume that the per-person generation rates will be fairly uniform and thus this source strength will be roughly proportional

Table 10.2.   Sources of Air Contaminants by Activity of Origin

Air contaminants from the occupants themselves
Air contaminants from construction or renovation
Air contaminants from building support activities
Air contaminants arising from maintenance of the HVAC system
Air contaminants arising from maintenance of the building

to the number of people present. A much greater individual variation can be expected for the other bioeffluents. Secondly, CO_2 concentrations can be measured with relative ease.

In terms of the transmission of "germs" (i.e., viruses and bacteria) from person to person, it needs to be remembered that there are three principle vectors, or modes, of transmission. One is nose-to-hand-to-hand-to-nose route, another is the oral-fecal route, and the third is the airborne nose-to-nose route. The importance of one versus the other will depend on the specific disease agent involved. There are, for instance, specific diseases that are referred to as "air-borne infections." One of these diseases, psittacosis or Parrot Fever, involves as its source a sick bird. Typical outbreaks have varied from homes with a sick parakeet to turkey-processing plants. Another airborne infection called Q fever caused a dramatic epidemic in Oakland, CA in 1959. As reported by Wellock,[4] a total of 75 cases were confirmed in persons who either resided in or were shown to have been exposed in a certain limited area about 7 miles long and $^1/_2$ to 1 mile wide. At the head of the area and in line with the prevailing winds was a rendering plant located where sheep and goats were processed.

Of more specific concern these days is the possible transmission of tuberculosis (TB). In fact, one of my more specialized tracer testing efforts has been to evaluate the effective ventilation rates (EVRs) in clinics where aerosolized pentamidine is being administered to AIDS patients in treatment rooms. Since the administration of this drug causes a lot of coughing, the staff involved requires adequate protection against TB.

In the more typical situations involving the transmission of colds and flu, the specific mechanism may be less clear. Often, the complaint is heard as part of a building investigation that "when one person comes in with a cold, it seems like everyone else gets it." It is true that in a situation like this with a contagious individual, the less ventilation that is being provided, the greater the likelihood that other people will receive a sufficiently high dose that they too will contract the illness. It is also true that the nose-to-hand-to-hand-to-nose route of transmission may be largely responsible for this. I have heard about an experiment at some university where a group of students with colds were invited to play some games of cards. The deck of cards they played with was then brought across campus to a group of healthy students. After this second group played with this particular deck of cards, they too came down with colds.

In addition to being sources of disease agents, the activities of the people in the building, such as smoking, sweating, cooking, or wearing perfumes, can all be considered as potential sources of air contaminants. For instance, if people are permitted to indulge their nicotine addiction and smoke in a building, this uncontrolled combustion process represents a source of over 4000 different chemical compounds, some of which have been identified as carcinogenic. Smoking areas, if they exist, need to be separately ventilated, negatively pressurized in relation to surrounding interior spaces, and supplied with higher ventilation rates than that provided in non-smoking areas. It is also recommended that the air leaving a smoking area should be exhausted directly outdoors and not recirculated within the building. The rationale for these actions was pointed out in a 1986 report of the Surgeon General; *The Health Consequences of Involuntary Smoking* states that "the case against involuntary smoking is more than sufficient to justify appropriate remedial action to protect the non-smoker from environmental tobacco smoke."

Air Contaminants from Construction or Renovation

During construction or renovation, numerous materials are used that have the potential to be sources of air contaminants. The interior building components and furnishings can release chemicals from their manufacture. This includes both VOCs as well as inorganic chemical compounds. This situation will be most severe in association with renovation activities. First, there are the construction-related sources: the generation of dust and fibers from demolition. This can then be followed by the chemical compounds such as joint compound dust, adhesives, caulks, and paints. This is typically followed by the introduction of new materials, carpets and partitions, which will have their highest levels of emissions when brand new.

Air Contaminants from Building Support Activities

Included in this category would be the office equipment (such as copiers and laser printers) that can emit VOCs or ozone. The use of supplies for this office equipment, such as solvents, toners, or ammonia (from blue print machines), should also be reviewed as potential sources of contaminants. Special activities (such as shops, photographic dark rooms, labs, or cleaning processes) can also be sources of contaminants.

This category of potential sources also includes activities or processes that create known air pollutants that are being exhausted directly to the outdoors. These sources (such as kitchen exhausts or laboratory fume hoods) can degrade IAQ either by failing to be captured effectively or by having these emissions reenter the building by failing to adequately discharge them away from the air intakes.

Air Contaminants Arising from Maintenance of the HVAC System

The normal maintenance activities performed on the HVAC system can also be a source of air contaminants in the building. The cleaning of any part of the HVAC system should be scheduled during periods when the building is unoccupied in order to prevent exposure to chemicals and loosened particles. The AHU should not be used during the cleaning of ductwork or as an air movement device for this cleaning process. If biocides are used in an HVAC system, the only products acceptable for this purpose need to be registered by the EPA for such use. It should be noted, however, that one of the most effective biocides is merely a solution of household bleach. These products must be used only according the manufacturer's directions, with careful attention paid to the method of application. At present, EPA accepts claims and therefore registers antimicrobials for use only as sanitizers, not disinfectants or sterilizers in HVAC systems. Once cleaned, the HVAC components should be thoroughly rinsed and dried to prevent exposure of building occupants to the cleaning chemicals.

Air Contaminants Arising from Maintenance of the Building

Housekeeping activities are an important activity required for the maintenance of good IAQ. Deficiencies in this area can therefore create IAQ problems. Insufficient vacuuming by housekeeping services can permit the build-up of micropaper particles and dust debris which can then lead to complaints of skin irritation among the occupants in an area. This situation can lead people to believe that they have an infestation of "paper fleas." As stated elsewhere, the Entomology Office of the Harvard University, Department of Environmental Health and Safety, however, holds that there is no such insect as a paper flea. In another building where there were complaints of skin irritation, there were initial fears of the presence of an insect infestation of some kind. However, after repeated fumigations failed to reduce the complaints, another consultant was brought in who correctly determined that the skin irritation was being caused by the combination of low humidities and the presence of fine fiberglass fibers introduced into the supply air from insulation around the perimeter of the return plenum.

In addition to insufficient housekeeping leading to IAQ problems, there was also a case of the actions of overzealous maintenance personnel leading to an IAQ problem. It has been reported by Kreiss et al.[5] that dried detergent residue left in carpets after they were shampooed with underdiluted carpet shampoo caused respiratory irritation among most employees in an office building and among all staff members and most children in a day-care center. The symptoms reported included cough, dry throat, difficulty in breathing, nasal congestion, and headache. Eye irritation was also noted by the day-care center staff members. Symptoms persisted for many weeks until the carpets were wet extracted. The major ingredient of the three shampoo products implicated in these two outbreaks and in a third similar report is sodium dodecyl sulfate, a respiratory irritant in mice. Detergent dust is therefore to be considered as an example of an indoor air pollutant, especially when patients or employees complain of building-specific respiratory or eye irritation.

SOURCES OF AIR CONTAMINANTS BY THEIR PROPERTIES

The basic categories of physical properties that can be used to distinguish among various air contaminants can be summarized as follows:

- Viable particles: viruses, bacteria, and spores
- Nonviable particles: dusts, droplets, and fibers
- Gases: VOCs, ozone, and carbon monoxide

This distinction is not absolute, however, because air contaminants may evaporate from being liquid droplets to become gases. Indoor air often contains a variety of contaminants at concentrations that are far below any standards or guidelines for occupational exposure. Given our present knowledge, it is difficult to relate complaints of specific health effects to exposures to specific pollutant concentrations, especially since the significant exposures may be to low levels of mixtures of air pollutants. One generalization that can be made, however, is that if people are complaining of odors, there are only two potential sources: one is from chemicals, while the other is from sites of microbiological growth.

Viable Particles

Viable particles, because they are capable of reproducing themselves, have the potential to increase their concentrations over time. This capability is very different from air contaminants characterized as physical or chemical agents which cannot increase their concentrations unless there is more source material available. Microbiologic contaminants come in a wide variety of forms. Outdoors there are pollens from grasses, trees, and other plants that contribute to allergic reactions in sensitive individuals. Microbiological contaminants in the indoor environment can also include microbial cells such as viruses, bacteria, as well as fungal spores, protozoans, algae, animal dander and excreta, plus insect excreta and fragments.

Nonviable Particles

Nonviable particles can arise from physical processes such as crushing, grinding, or as products of incomplete combustion, such as the respirable particulate from smoking. Other physical properties include the wearing away of rubber, such as from fan belts in the HVAC equipment or from tire wear on roadways in proximity to outdoor air intakes. Moving air can also entrain particles into the air stream. Examples include fiberglass eroded from duct liners or dirt deposits above suspended ceilings that gets introduced into the supply air via fan-power mixing boxes. With respect to the HVAC system, the quality of the filters and their installation will play a major role in the amount of particulate matter delivered with or removed from the supply air stream. In the occupied space itself, in addition to smoking, copiers can be the source of toner particles and paper fibers.

Gases

With respect to gases, copiers can be a source of ozone or plasticizer fumes which are generated as the toner matrix is fused to the paper. Other gases can be categorized as being products of combustion. These combustion products include the inorganic gases such as NO, NO_2, CO, and CO_2, as well as particles. For organic gases (or VOCs as they are also called) the sources include materials of construction, furnishings, and consumer and/or office products. In a recent Environmental Protection Agency study of air quality in 10 public access buildings, more than 500 VOCs were identified.[6] This magnitude of the number of compounds present is indicative of the organic stew of VOCs that can accumulate in the indoor environment as compared with the outdoors, and the need for adequate and effective ventilation to dilute and remove them.

REFERENCES

1. Hujanen, M., O. Seppänen, and P. Pasanen. 1991. "Odor Emission from the Used Filters of Air-Handling Units." *ASHRAE IAQ '91, Healthy Buildings.* pp. 329–333.
2. Air-Conditioning and Refrigeration Institute (ARI). *Air Conditioning and Refrigerating Equipment General Maintenance Guidelines for Improving Indoor Air Environment.* Arlington, VA. 8 pp.
3. Morey, P. R. and C. M. Williams. 1991. "Is Porous Insulation Inside an HVAC System Compatible with a Healthy Building?" *ASHRAE IAQ '91, Healthy Buildings.* pp. 128–135.
4. Wellcock, C. E. 1960. "Epidemiology of Q fever in the urban East Bay area." *California's Health,* 18:73.
5. Kreiss, K., M. G. Gonzalez, K. L. Conright, and A. R. Scheere. 1982. "Respiratory Irritation Due to Carpet Shampoo: Two Outbreaks." *Environ. Int.,* Vol 8:337–341.
6. Sheldon, L. S. et al. 1988. *Indoor Air Quality in Public Buildings, Vol. 1: Project Summary.* U.S. Environmental Protection Agency. Washington, D.C. Publication No. EPA/600/S6-88/009a.

GLOSSARY and ACRONYMS

AC/min — Air changes per minute.

ACH — Air changes per hour. A measurement of the ventilation or supply air rate calculated by dividing the volume of air delivered by the volume of the space receiving the air.

Air cleaning — An IAQ control strategy to remove various airborne particulates and/or gases from the air.

AHU — Air handling unit. A component of an HVAC system that includes the fan or fans, filters and coils to condition air.

ASHRAE — American Society of Heating, Refrigerating, and Air-Conditioning Engineers, Inc.

Biological contaminants — Agents derived from or that are living organisms (e.g. viruses, bacteria, fungi, and mammalian and bird antigens) that can be inhaled and can cause many types of health effects including allergic reactions, respiratory disorders, hypersensitivity diseases, and infectious diseases.

Breathing zone — Area of a room in which occupants breathe as they stand, sit, or lie down. Sometimes referred to as the portion of a space from 3" to 72" above the floor.

Building envelope — Elements of the building, including all external building materials, windows, and walls that enclose the internal space.

BRI — Building relating illness. Diagnosable illness whose symptoms can be identified and whose cause can be directly attributed to airborne building pollutants (e.g. Legionnaire's disease, hypersensitivity pneumonitis).

CAV — see Constant air volume.

CFM — Cubic feet per minute.

CO — Carbon monoxide.

CO_2 — Carbon dioxide.

Constant air volume — Type of HVAC system that varies the temperature, but not the volume, of delivered supply air to regulate thermal conditions.

DA — Distribution apportionment. The relationship between the proportion of the OA quantity being delivered to portion a building and the proportion of the people in the building that are actually located in that portion of the building.

DI — Distribution integrity. The relationship between the OA quantity entering the HVAC equipment and the OA that actually gets delivered to the building occupants.

DDC — Direct digital control. Type of control equipment for HVAC systems.

ETS — Environmental tobacco smoke.

EVR — Effective ventilation rate. The ventilation rate based on the actual quantity of outdoor air delivered to the occupied areas of a building or space.

ft^2 — Square feet.

HVAC — Heating, ventilating and air conditioning system.

IAQ — Indoor air quality.

OA — Outdoor Air.

OASI — Outdoor air supply index.

PA — Pascal. Unit of pressure measurement

PIU — Perimeter induction unit.

ppb — Parts per billion.

PPM — Parts per million.

RA — Return air.

Reentrainment — Situation that occurs when a portion the air being exhausted from a building is brought back into the building, either through the OA intake of the HVAC system or some other opening in the building envelope.

RF — Radio frequency. Portion of electromagnetic spectrum.

RTU — Roof top unit. A packaged AHU located on the roof.

SA — Supply air.

Sick building syndrome — Term sometimes used to describe situations in which building occupants experience acute health and/or comfort effects that appear to be linked to time spent in a particular building, but where no specific illness or cause can be identified. The complaints may be localized in a particular room or zone, or may be spread throughout the building.

Short-circuiting — Situation that occurs when the supply air flow to exhaust registers before entering the breathing zone.

SF_6 — Sulfur hexafluoride. A physiologically inert gas used as a tracer in building investigations.

Stack Effect — Pressure-driven airflow produced by convection as heated air rises, creating a positive pressure area at the top of a building and a negative pressure area at the bottom of the building. The stack effect can overpower the mechanical system and disrupt ventilation and circulation in a building.

VAV — Variable air volume. Type of HVAC system that varies the volume of delivered supply air to regulate thermal conditions.

Ventilation air — Outdoor air delivered to occupied spaces of a building.

VOC — Volatile organic compounds. Compounds that evaporate from the many housekeeping, maintenance, and building materials made with organic chemicals. Biological organisms can also be a source of VOCs. In sufficient quantities, VOCs can cause eye, nose and throat irritations, headaches, dizziness, visual disorders, and memory impairment.

Index